ATP 3-05.1

Unconventional Warfare

September 2013

DISTRIBUTION NOTICE: Distribution authorized to U.S. Government agencies and their contractors only to protect technical or operational information from automatic dissemination under the International Exchange Program or by other means. This determination was made on 24 July 2013. Other requests for this document must be referred to Commander, United States Army John F. Kennedy Special Warfare Center and School, ATTN: AOJK-CDI-SFD, 3004 Ardennes Street, Stop A, Fort Bragg, NC 28310-9610.

DESTRUCTION NOTICE: Destroy by any method that will prevent disclosure of contents or reconstruction of the document.

FOREIGN DISCLOSURE RESTRICTION (FD 6): This publication has been reviewed by the product developers in coordination with the United States Army John F. Kennedy Special Warfare Center and School foreign disclosure authority. This product is releasable to students from foreign countries on a case-by-case basis only.

Headquarters, Department of the Army

ATP 3-05.1

Army Techniques Publication
No. 3-05.1

Headquarters
Department of the Army
Washington, DC, 6 September 2013

Unconventional Warfare

Contents

		Page
	PREFACE	iv
	INTRODUCTION	v
Chapter 1	**OVERVIEW**	**1-1**
	Description	1-1
	The 21st Century Strategic Environment	1-2
	National Strategic Policy Option	1-2
	The Military Instrument of National Power	1-3
	Ways the United States Conducts Unconventional Warfare	1-3
	Resistance and Insurgent Movement Characteristics	1-5
	Feasibility for United States Sponsorship	1-6
Chapter 2	**FUNDAMENTALS**	**2-1**
	Why and How Populations Resist	2-1
	The Seven Phases of Unconventional Warfare	2-8
	The Components of an Insurgency	2-16
	Additional Components of an Insurgency	2-20
Chapter 3	**ACTIVITIES**	**3-1**
	Preparation of the Environment	3-1
	Subversion	3-2
	Information Activities	3-3
	The Subversive Manipulation of Crowds, Riots, and Demonstrations	3-10
	Sabotage	3-14
	Nonconventional Assisted Recovery	3-18

DISTRIBUTION RESTRICTION: Distribution authorized to U.S. Government agencies and their contractors only to protect technical or operational information from automatic dissemination under the International Exchange Program or by other means. This determination was made on 24 July 2013. Other requests for this document must be referred to Commander, United States Army John F. Kennedy Special Warfare Center and School, ATTN: AOJK-CDI-SFD, 3004 Ardennes Street, Stop A, Fort Bragg, NC 28310-9610.

DESTRUCTION NOTICE: Destroy by any method that will prevent disclosure of contents or reconstruction of the document.

FOREIGN DISCLOSURE RESTRICTION (FD 6): This publication has been reviewed by the product developers in coordination with the United States Army John F. Kennedy Special Warfare Center and School foreign disclosure authority. This product is releasable to students from foreign countries on a case-by-case basis only.

	Guerrilla Warfare 3-20
	Intelligence Operations 3-38
Chapter 4	**SUPPORTING ACTIVITIES AND LEGAL CONSIDERATIONS 4-1**
	Interagency Support... 4-1
	Conventional Forces Support 4-2
	Military Information Support Operations 4-2
	Civil Affairs Operations 4-2
	Logistic Considerations.. 4-4
	Legal Considerations 4-11
Chapter 5	**CAMPAIGN PLANNING... 5-1**
	Unconventional Warfare Planning Considerations 5-1
	Joint Operation Planning Process 5-3
Appendix A	**CONSOLIDATED LIST OF UNCONVENTIONAL WARFARE COMPONENT CAPABILITIES... A-1**
Appendix B	**A 21ST CENTURY LAWRENCE OF ARABIA: ATTRIBUTES OF THE SPECIAL FORCES SOLDIER B-1**
Appendix C	**SELECTED EXAMPLES OF U.S.-SPONSORED UNCONVENTIONAL WARFARE ACTIVITIES.. C-1**
Appendix D	**NOTABLE INTERNATIONAL PRACTITIONERS OF INSURGENT, RESISTANCE, OR REVOLUTIONARY WARFARE THROUGHOUT HISTORY D-1**
Appendix E	**REVOLUTIONARY AND INSURGENT WARFARE ANALYSIS...................... E-1**
Appendix F	**UNCONVENTIONAL WARFARE CAMPAIGN ANALYSIS F-1**
	GLOSSARY Glossary-1
	REFERENCES References-1
	INDEX Index-1

Figures

Figure 2-1. Structure of an insurgency or resistance movement.. 2-3
Figure 2-2. The operations cell 2-17
Figure 2-3. The intelligence cell.. 2-17
Figure 2-4. The auxiliary cell... 2-18
Figure 2-5. Parallel cells 2-19
Figure 2-6. Cells in series 2-20
Figure 2-7. Joint special operations area "Debra" area complex sketch 2-22
Figure 3-1. Movement to an objective 3-25
Figure 3-2. Withdrawal from an objective 3-26
Figure 5-1. Relationship of end state (mission), objectives, effects, and tasks..................... 5-7
Figure 5-2. Sample line of operation for unconventional warfare 5-13
Figure 5-3. Sample lines of effort for unconventional warfare 5-14

Figure 5-4. Course of action decision brief 5-22
Figure 5-5. Commander's visualization 5-23
Figure 5-6. Joint Operation Planning and Execution System phasing model 5-24
Figure 5-7. Information operations/Military Information Support Operations concept of support plan format 5-26
Figure 5-8. Civil-military operations concept of support plan format 5-27
Figure A-1. Army component unconventional warfare capabilities A-1
Figure A-2. Army component derivative unconventional warfare capabilities A-2
Figure A-3. Navy component unconventional warfare capabilities A-3
Figure A-4. Navy component derivative unconventional warfare capabilities A-3
Figure A-5. Air component unconventional warfare capabilities A-4
Figure A-6. Air component derivative unconventional warfare capabilities A-4
Figure A-7. Marine component unconventional warfare capabilities A-5

Tables

Introductory Table 1. New Army terms vi

Preface

Army Techniques Publication (ATP) 3-05.1, *Unconventional Warfare*, provides the current United States (U.S.) Army Special Forces (SF) concept of planning and conducting unconventional warfare (UW) operations. ATP 3-05.1 describes UW fundamentals, activities, and considerations involved in the planning and execution of UW throughout the full range of military operations, and emphasizes UW as a strategic policy option. This publication serves as the doctrinal foundation for subordinate Army special operations forces (ARSOF) UW doctrine, force integration, materiel acquisition, professional education, and individual and unit training. This publication also serves as the Army's description of UW, which will be useful in the larger joint and interagency environment.

The principal audience for ATP 3-05.1 is commanders, staff officers, and operational personnel at the SF group (airborne) and theater special operations command (TSOC) levels. This publication is specifically for SF Soldiers; however, it is also intended for Army-wide use to improve the integration of SF into the plans and operations of other special operations forces (SOF), conventional military forces, and interagency and intergovernmental activities.

Commanders, staffs, and subordinates ensure their decisions and actions comply with applicable U.S., international, and, in some cases, host-nation laws and regulations. Commanders at all levels ensure their Soldiers operate in accordance with (IAW) the law of war and the rules of engagement. (See Field Manual [FM] 27-10, *The Law of Land Warfare*.)

ATP 3-05.1 uses joint terms where applicable. Selected joint and Army terms and definitions appear in both the glossary and the text. Terms for which ATP 3-05.1 is the proponent publication (the authority) are marked with an asterisk (*) in the glossary. Definitions for which ATP 3-05.1 is the proponent publication are boldfaced in the text. These terms and their definitions will be in the next revision of Army Doctrine Reference Publication (ADRP) 1-02, *Operational Terms and Military Symbols*. For other definitions shown in the text, the term is italicized and the number of the proponent publication follows the definition.

ATP 3-05.1 applies to the Active Army, the Army National Guard/National Guard of the United States, and the United States Army Reserve unless otherwise stated.

The proponent of ATP 3-05.1 is the Special Operations Center of Excellence. The preparing agency is the Special Forces Doctrine Division, Capabilities Development and Integration Directorate, United States Army John F. Kennedy Special Warfare Center and School (USAJFKSWCS). Send comments and recommendations on Department of the Army (DA) Form 2028 (Recommended Changes to Publications and Blank Forms) to Commander, USAJFKSWCS, ATTN: AOJK-CDI-SFD, 3004 Ardennes Street, Stop A, Fort Bragg, NC 28310-9610; by e-mail to AOJK-DT-SF@soc.mil; or by submitting an electronic DA Form 2028.

Introduction

ATP 3-05.1, *Unconventional Warfare*, is the Army's doctrinal foundation for UW and is the broadest and most comprehensive United States Government (USG) doctrinal publication on the subject of UW. ATP 3-05.1 therefore provides doctrine directly useful to all users within the U.S. Army, but is deliberately intended to be useful to other Services in the Department of Defense (DOD) and joint, interagency, intergovernmental, and multinational (JIIM) audiences. Moreover, although UW is inherently a sensitive subject, ATP 3-05.1 is intentionally kept unclassified to make it accessible to civilian policy makers with a role in oversight and support of UW activities.

ATP 3-05.1 is written to emphasize the strategic and operational utility of UW as a policy option available to national-level and theater-level decisionmakers. The ATP is therefore written for planners at the TSOC and SF group level who would be charged with recommending and planning strategic and operational options to geographic combatant commanders (GCCs), Ambassadors, and interagency decisionmakers at all levels of the USG. ATP 3-05.1 contains five chapters and six appendices that are summarized in the following paragraphs.

Chapter 1 provides a conceptual overview of UW, sets it within the context of national policy, and discusses why UW is a valuable policy option. The chapter provides an appreciation for the criticality of considering when and to what degree UW is feasible.

Chapter 2 considers how and why populations resist. It outlines the seven-phase model of U.S.-sponsored insurgency and explains the classic and affiliated components of insurgent and resistance organizations.

Chapter 3 is a detailed discussion of the activities that comprise UW: preparation of the environment, subversion, sabotage, unconventional assisted recovery, guerrilla warfare, and intelligence operations. Although these subjects have always been referred to in UW publications, detailed discussion of them has been reduced in Army SOF doctrine since the end of the Cold War. This chapter reintroduces a detailed discussion of this content.

Chapter 4 provides practical information on supporting activities for UW, with a significantly expanded section on logistical support and sustainment planning.

Chapter 5 outlines a procedure for UW campaign planning. This is a new feature in UW doctrine intended to reinforce the strategic and operational utility of UW, and contribute to ATP 3-05.1 being a single-volume resource for UW campaign planners.

Appendix A is an unprecedented list of joint SOF component capabilities for UW derived from the United States Special Operations Command (USSOCOM) Directive 525-89, *(S/NF) Unconventional Warfare (U)*. The list adds conceptual clarity to Service roles in UW useful to any joint and interagency planning group.

Appendix B provides a list of characteristic attributes for those most likely to succeed in executing UW based on the Army SF Soldier.

Appendix C presents examples of previous U.S.-sponsored UW useful to UW campaign planners' design process.

Appendix D presents examples of previous international practitioners of insurgent or resistance warfare useful to UW campaign planners' design process.

Appendix E analyzes the characteristics of significant recent revolutionary and insurgent warfare movements that provide instructive examples of modern irregular warfare.

Appendix F is a comprehensive analytical tool for strategic and operational analysis of potential UW target countries.

Introduction

Based on current doctrinal changes, certain terms for which ATP 3-05.1 is proponent have been added, rescinded, or modified for purposes of this publication. The glossary contains acronyms and defined terms. ATP 3-05.1 added the terms listed in Introductory Table-1.

Introductory Table 1. New Army terms

Term	Remarks
area command	New term and definition.
government-in-exile	New term and definition.
guerrilla	New term and definition.
guerrilla base	New term and definition.
mission support site	New term and definition.
pilot team	New term and definition.
shadow government	New term and definition.

Chapter 1
Overview

> *Unconventional warfare:* Activities conducted to enable a resistance movement or insurgency to coerce, disrupt, or overthrow a government or occupying power by operating through or with an underground, auxiliary, and guerrilla force in a denied area.
>
> Joint Publication (JP) 3-05, *Special Operations*

National decisionmakers can choose to employ special operations across the range of military operations at all levels of war and throughout all phases of a campaign or operation. UW is one such special operation particularly useful when discrete, indirect, and relatively small-scale options that leverage indigenous actors are desirable. This publication describes the fundamentals and activities of what UW is, provides examples of the insurgent and revolutionary environment in which UW is conducted, and identifies participants and planning procedures useful to UW campaign planning.

DESCRIPTION

1-1. UW consists of activities conducted to enable a resistance movement or insurgency. UW supports the efforts of an opposition group (resistance movement or insurgency) in conflict with an established government, governing authority, or occupying power exercising authority and/or control over a specific territory and/or population.

1-2. UW aims to coerce, disrupt, or overthrow an enemy occupying power or government. The objectives of UW span from the provision of support to a group resisting the rule of a governing authority, to the installation of another government. A government need not be a national government recognized by the international community, but it must exercise public policy and provide state-like functions for a population. Other examples of a governing authority may include a tribal government, a warlord's rule, or a malign actor exercising de facto authority (like the Taliban before September 2001) over a specific sub-state territory or population.

1-3. UW is conducted by operating through or with an underground, auxiliary, and guerrilla force. Opposition groups organize around these interrelated elements based on their environment (urban, rural, or mixed), the governing authority's capabilities, and the capabilities inherent in the organization. If an insurgency or resistance is being supported by the USG, SOF may operate with or through all elements of the opposition group, or be tasked to develop any missing elements to ensure the survivability and success of the organization. Additionally, SOF might potentially execute UW activities with, through, or in support of a third party or a friendly nation-state. UW's focus is on applying USG capabilities indirectly through indigenous actors.

1-4. UW is conducted in a denied area. The initial UW actions and activities in support of an opposition group, or targeted against a governing authority, are typically conducted in areas and environments where the governing authority has the capacity and will to deny overt freedom of action to the opposition group. However, not all actions and activities supporting UW are conducted in such environments. UW may be conducted or supported from artificial, physical, or virtual safe havens within or outside the joint special operations area (JSOA).

Chapter 1

THE 21ST CENTURY STRATEGIC ENVIRONMENT

1-5. The increasing complexity of future operational environments and the resulting diffusion of power from nation-states to nonconventional, hybrid, irregular, nonstate, or trans-state actors may render traditional applications of the elements of national power less effective or less appropriate for achieving U.S. national objectives.

1-6. The beginning of the 21st century has been focused on disrupting global terrorist networks—a task that will continue for the foreseeable future. While UW would not be waged against terrorist groups themselves, it could conceivably be used against state sponsors of terrorists or third-party nations that provide terrorists sanctuary, transit, or other support, in an effort to coerce these nations to stop their support, or to disrupt these nations and their support to terrorist groups. Policy makers, commanders, and planners must remember that UW does not automatically mean and rarely equates to "overthrow."

1-7. However, the rise and fall of states is a constant throughout history. It is virtually certain that some of those rising powers will have interests that are hostile to those of the United States. Such powers may develop into legitimate targets for the coercion and disruption options of national statecraft provided by U.S. UW capabilities. And although rare, UW is still valid in the 21st century to enable a resistance to fight an occupying power, or even to enable an insurgency to pursue regime change in a hostile state.

NATIONAL STRATEGIC POLICY OPTION

1-8. UW is a national strategic politico-military tool. Under certain circumstances, UW may provide the only feasible option by which the USG can achieve its objectives—whether those objectives are to influence, coerce, disrupt, or replace a governing authority; shape attitudes and influence the behavior of a relevant population, group, or individual; or counter an irregular threat. UW is essentially a combination of direct and indirect applications of national power—one that leverages relevant foreign opposition groups, insurgencies, or resistance movements opposing a national government or other governing authority to act in support of the sponsoring state's national objectives.

1-9. 21st century UW translates national strategy and policy into an operational concept that provides national decisionmakers an appropriate, and possibly cost-effective, strategic policy option composed of a series of interrelated USG actions aimed at achieving the strategic and operational objectives within the (potentially distributed) JSOA. Such activity is not without political risk, but such risk could potentially be less than the risk associated with inaction on the part of the USG or the execution of a conventional military campaign.

1-10. Establishing the necessary conditions to successfully execute UW involves significant strategic political risk. The sensitive nature of special operations, the methods by which they are conducted, and the ramifications of failure require the joint force commander (JFC) to conduct extensive planning and preparation across the DOD to reduce risk to an acceptable level. Equally important as the mitigation of strategic risk is the implied requirement for extensive coordination, deconfliction, and integration across the USG and its partners. Potential strategic risk and political sensitivity of UW operations require planners to approach UW as a whole-of-government campaign. Responsibilities should be assigned IAW each organization's core competencies.

1-11. UW, as an element of irregular warfare, has applicability across the range of military operations in various operational environments. Opposition groups may develop anywhere to challenge the government or occupying power when there exists dissatisfaction or grievances on the part of a portion of the population. No nation or geographical region is immune to the possibility. If these groups oppose government or occupying powers that are hostile to U.S. interests and they have the potential or are beginning to transition to organized violence, UW may be a suitable option vice conventional military actions. UW provides a way for the USG to influence the actions, attitudes, and policies of the relevant actors and shape the competition for legitimacy.

1-12. There are certain prerequisite conditions necessary for successful UW, some of which are beyond control and others that can be influenced. The most critical condition is the compatibility of the opposition group's objectives and those of the USG (aligned or tangentially aligned). Other prerequisite conditions include vulnerability of the governing authority's legitimacy, assets, infrastructure, and ability to control its population and territory; viable indigenous opposition group leadership (not transplanted surrogates); willingness of the opposition group to partner with the identified sponsor; and an environment suitable for UW.

THE MILITARY INSTRUMENT OF NATIONAL POWER

1-13. The optimal conduct of UW uses a comprehensive approach where military operations support, and/or are supported by, other USG departments and agencies as part of a whole-of-government campaign. However, under Title 10, United States Code (USC), *Armed Forces*, Section 167, *Unified Combatant Command for Special Operations Forces*, the USSOCOM Commander has a responsibility to be prepared to conduct a UW campaign without the participation of some or any of the other USG departments and agencies, including the necessity to conduct inherently civilian roles and functions.

1-14. UW requires a campaign mindset and, like any 21st century military venture, requires early and continuous interagency engagement, collaboration, and an in-depth common understanding of the strategic conduct of UW (Chapter 5 includes more information). Acting as partners, USG departments and agencies must embrace ever closer integration, collaboration, and innovation to facilitate support to the opposition group's diplomatic/political, informational, military, and economic efforts in pursuit of objectives that align with U.S. national interests. UW campaign plans that originate from this type of collaborative and effective interagency partnership and planning directly impact the efficiency and success of the collective effort.

1-15. The potential for an adversary's critical assets and activities to be integrated within global or regional systems reinforces the requirement for the joint force to collaboratively develop a highly distributed UW effort involving multiple lines of operations, and selectively apply the various authorities and capabilities resident within the USG and its partners. Such collaboration significantly enhances the USG's ability to achieve its national objectives and to establish conditions for postconflict support to the resultant governing authority.

1-16. The USSOCOM UW capabilities satisfy GCCs' campaign requirements through the provision of special operations forces' support to insurgencies and resistance movements, from initial contact through transition to a legitimate governing authority in denied and uncertain environments. USSOCOM has a Title 10, USC, Section 167, responsibility to provide specialized forces developed, recruited, selected, trained, organized, equipped, and capable of conducting the military aspects of UW.

1-17. The United States Army Special Operations Command (USASOC) serves as USSOCOM's lead component in all matters pertaining to SOF conduct of UW and executes UW as a primary core task. Naval Special Warfare Command (NAVSPECWARCOM), Marine Corps Forces Special Operations Command (MARSOC), and Air Force Special Operations Command (AFSOC) all serve as USSOCOM supporting components for UW and support UW as a secondary task. Appendix A contains an itemized list of all primary and supporting capabilities required of each DOD component Service as listed in USSOCOM Directive 525-89. Appendix B discusses attributes expected of a professional unconventional warrior.

WAYS THE UNITED STATES CONDUCTS UNCONVENTIONAL WARFARE

1-18. There are two distinct types of UW efforts, each with distinct levels of associated political risk. One type is when UW is only one line of operation within the military effort and the military instrument of national power is the dominant effort within the larger context of USG unified action. In this type of campaign, military involvement is large-scale and overt, and typically involves major operations.

1-19. The second type is when UW is employed as the main effort, either as an initiative or as a response to aggression. In this type of campaign, overt and dominant military force is impractical or impossible. Although this type of campaign may represent the nation's main effort, the military instrument of power is likely to be in a supporting role to the other instruments. UW is particularly useful in this kind of campaign; it is a small-scale, discrete—or even nonattributable—method that relies on primarily indirect applications of national power.

Chapter 1

U.S. MILITARY LARGE-SCALE INVOLVEMENT SCENARIOS

1-20. There are three possible goals of large-scale involvement. The goal is to facilitate the eventual introduction of conventional forces, to facilitate friendly offensive or defensive operations, or to divert enemy resources away from other parts of the operational area.

1-21. UW forces can function as effective instruments in the psychological preparation of the population for the introduction of conventional forces. Furthermore, deception and other measures can convince enemy leaders to divert resources away from the main area of effort when it is not necessary to do so. For example, the United States can deliver messages suggesting guerrilla operations will occur in certain locations, causing enemy leaders to divert their forces away from the actual route of advance to meet a nonexistent threat. Examples of this type of UW effort by the United States include the following:

- European and Pacific Theaters (1942–1945).
- North Korea (1951–1953).
- Cold War Contingency Plans for Eastern Europe (1952–1989).
- Kuwait (1990–1991) (aborted).
- Afghanistan (2001–2002).
- Iraq (2002–2003).

1-22. During large-scale UW, operations focus largely on military aspects of the conflict because of the eventual introduction of conventional forces. The task is normally to disrupt or degrade enemy military capabilities in order to make them more vulnerable to the pending introduction of conventional invasion forces. The USG can use actions and messages to disrupt and degrade enemy capabilities by lowering their morale and unit cohesion. This can increase desertion, surrender, and malingering among their ranks. Use of such techniques increases the potential for enemy unit breakdown to the point of rendering them combat-ineffective.

1-23. Resistance forces assume a one-time greater degree of risk in large-scale involvement scenarios by exposing almost their entire infrastructure in exchange for the possibility of success and linkup with friendly coalition forces following an invasion. The ultimate challenge is synchronizing resistance efforts while maintaining a degree of operational security for the invasion.

1-24. If the intent of the UW operation is to develop an area in order to facilitate the entry of an invasion force, the challenge is to ensure that the operations of the resistance complement (rather than inadvertently interfere with or even compromise) those of the invasion forces. If the timing is wrong or the conventional invasion forces fail to liberate the territory and link up with resistance forces, it is likely that the resistance organization (guerrillas, underground, and auxiliary personnel) will suffer significant losses.

1-25. With a few exceptions, it is relatively simple for U.S. forces to compel an adversary to commit forces to an area away from a possible invasion site. The challenge in this scenario is determining which resistance actions trigger the desirable responses and when to begin those operations to appropriately affect the adversary's decision cycle. If U.S. forces do not coordinate these operations with the invasion force or do not time the operations orrectly, they can cause significant negative consequences.

U.S. MILITARY SMALL-SCALE INVOLVEMENT SCENARIOS

1-26. In general, the United States uses limited-involvement operations to pressure an adversary. Examples of this type of UW effort by the United States include the following:

- The Baltic States (Estonia, Lithuania, Latvia) (1950s).
- Guatemala (1954).
- Albania (1949–1954).
- Tibet (1955–1965).
- Indonesia (1958).
- Cuba (1960s).
- Laos (1960–1962/1962–1970).

- North Vietnam (1961–1964) (aborted).
- Afghanistan (1980s).
- Nicaragua (1980s).

1-27. During limited-military-involvement missions, the overall operation takes place in the absence of overt or eventual hostilities from the sponsor. Such operations take on a strategic and sensitive political aspect.

1-28. Typically, the United States limits its direct military involvement, which mitigates the risks of unintended consequences or premature escalation of the conflict. During limited-military-involvement operations, the manner in which forces operate significantly differs from that of large-scale military involvement scenarios. Without the benefit of a conventional invasion force, the resistance forces must limit overt exposure of their forces and supporting infrastructure in order to sustain operations over a protracted period. Moreover, although military forces employed may be few in number and discretely employed, other committed resources (such as insurgent funding or information operations [IO] campaigns) may be substantial. (Appendix C discusses more historical details on U.S.-sponsored UW activities.)

RESISTANCE AND INSURGENT MOVEMENT CHARACTERISTICS

1-29. There are certain characteristics of a resistance movement that make U.S. support favorable. This short, select list is different from the various lists of characteristics by which to analyze insurgencies and resistance movements. The following paragraphs discuss the characteristics of a favorable movement.

WILLINGNESS TO COOPERATE WITH THE UNITED STATES

1-30. A genuine willingness to collaborate and cooperate with the United States must exist within the leadership of the indigenous force. It is unrealistic to expect a leader to relinquish control of his forces to the United States. In general, insurgent leaders expect to retain authority and control over their forces while benefiting their cause by collaborating with the United States. Persuasive messages targeting key leaders and groups may increase their willingness to accept U.S. support.

COMPATIBLE OBJECTIVES AND IDEOLOGY

1-31. Successful movements must have compatible objectives and an ideology that binds their forces together. Organizations bound through some commitment other than common ideology—such as forced conscription or hired mercenaries—typically are only marginally capable over a protracted period. Armed groups may find a common bond in ethnicity, religion, or tribal ties. SOF units can use persuasive techniques and messages emphasizing commonalities to unite different groups for a common cause. Once the groups unite, other messages can reinforce unity by building morale, reinforcing organizational cohesion, and emphasizing mutual goals.

CAPABLE RESISTANCE LEADERSHIP

1-32. Resistance movement leaders are cautious of quickly forming new partnerships. In order to understand insurgent leaders, it is critical to understand their motivation and desires. Planners must consider what the United States is requesting and offering in return from the insurgent's perspective. The best leader is not always the one that is the easiest to work with initially. In fact, an overly accommodating leader could be a desperate and incapable leader primarily interested in personal gain. Similarly, a seemingly indifferent leader could be an effective leader that is unimpressed with offers of support without an assurance of long-term commitment because of the potential risk involved. The determination of the appropriateness of U.S. support requires an in-depth understanding of the resistance leadership and organization. This level of fidelity normally requires a degree of first-hand observation in order to develop an educated assessment.

1-33. Military Information Support Operations (MISO) assessments are important sources of information on the psychological characteristics of leaders and groups. This analysis provides a degree of prediction about the future behavior of these potential partners. With prediction comes a degree of confidence in knowing how potential resistance leaders will conduct themselves in the UW effort. It also provides information on guerrilla leader expectations for their forces in terms of the method of fighting, treatment of civilians, and other key aspects that can have political and legal ramifications for the operation. This information aids the commander or other decisionmakers in determining the appropriateness of any support to the movement.

FEASIBILITY FOR UNITED STATES SPONSORSHIP

1-34. Some of the key considerations for assessing UW feasibility, adequacy and appropriateness include the following:

- When properly coordinated and executed, UW may help set conditions for international crisis resolution on terms favorable to the United States or allies without the need for an overt U.S. conventional force commitment.
- The conduct of UW can have a strategic military-political utility that can alter the balance of power between sovereign states.
- UW often entails significant political risk both at home and abroad.
- Although UW can be a line of operation during wartime, many UW activities are conducted in peacetime and subordinate to the practices of normal statecraft.
- UW usually requires a varying mix of clandestine and covert activities.
- Each specific instance of UW is unique based on the specific human and environmental characteristics of the mission.
- UW operations, especially those in which insurgent or resistance underground networks have to be established, require time to mature and reach maximum effectiveness.

1-35. There are certain environments and situations that make UW the best option. Although outside forces could alter and shape the existing environment to some degree, they cannot artificially manufacture or transplant it. There are three criteria planners use when deciding to provide support. The first criterion is feasibility. Feasibility is dependent upon the physical and human conditions of the environment. The second criterion is adequacy. Adequacy, and the third criterion—acceptability—are dependent upon the characteristics of the resistance movement. U.S. UW forces possess capabilities that can profoundly affect the operational environment through shaping operations that influence behavior in support of U.S. objectives. They can also influence resistance movement characteristics, making them more appropriate to the mission. For example, U.S. UW forces could emphasize guerrilla adherence to international norms and standards of behavior. Alignment of the resistance movement's ideology with U.S. values and mores may also factor into the decision whether or not the United States will support a resistance effort.

PHYSICAL AND HUMAN ENVIRONMENTAL CONDITIONS

1-36. There are specific physical and environmental conditions that allow for a successful resistance or insurgency. The three main conditions are a weakened or unconsolidated government or occupying power, sufficient popular will to resist, and favorable terrain from which an element can organize and wage subversion and armed resistance.

Weakened or Unconsolidated Government or Occupying Power

1-37. Conditions must sufficiently divide or weaken the organizational mechanisms that the ruling regime uses to maintain control over the civilian population for the resistance to successfully organize the minimum core of clandestine activities. It is extremely difficult to organize successful resistance under a fully consolidated government or occupying power with a strong internal security apparatus. Despite the general dissatisfaction of the population, the resistance has little chance of developing the supporting infrastructure it needs to succeed. Planners need to recognize the significant differences in the ability of different elements to exert control over a population. A recent foreign occupier does not have the same ability as an indigenous long-standing dictatorial regime that has had years to consolidate power.

Will of the Population

1-38. A percentage of the population must possess not only the desire to resist but also the will to bear the significant hardships associated with repressive countermeasures by the government or occupying power. Populations that the regime subjugates or indoctrinates for long periods are less likely to possess the will required to sustain a prolonged and difficult struggle. Populations living under repressive conditions generally either retain their unique religious, cultural, and ethnic identity or begin to assimilate with the regime out of an instinct to survive. Planners need to distinguish between the population's moral opinion of their "oppressors" and their actual willingness to accept hardship and risk on behalf of their values and beliefs. Populations recently overtaken by an occupying military force have a very different character than those that have had to survive for decades under an oppressive regime.

1-39. Information activities that increase dissatisfaction with the hostile regime or occupier and portray the resistance as a viable alternative are important components of the resistance effort. These activities can increase support for the resistance through persuasive messages that generate sympathy among populations. In almost every scenario, resistance movements face a population with an active minority supporting the government and an equally small militant faction supporting the resistance movement. For the resistance to succeed, it must convince the uncommitted middle population, which includes passive supporters of both sides, to accept it as a legitimate entity. A passive population is sometimes all a well-supported insurgency needs to seize political power. As the level of support for the insurgency increases, the passive majority will decrease. In addition, UW planners should not overlook the important role that diasporas can play in building international sympathy for the resistance and reducing support to the governing regime from the relative freedom outside the target country.

Favorable Terrain and/or Environment

1-40. In order to conduct operations, in most cases, resistance forces require human and physical terrain that provides safe haven. This terrain must possess enough security for resistance members to train, organize, and recuperate. The resistance must locate safe havens in relatively inaccessible areas that restrict the ability of the governing power's military force to project power and exert control. Examples of favorable physical terrain include relatively inaccessible areas, such as mountains, jungles, and swamps, which are restrictive to pursuing counterinsurgent forces. Artificial safe havens exist where the shielding of the resistance force is based on factors other than topographical difficulty. Resistance forces can sometimes find safe haven across international borders. Demographic groups within the country which the government security forces are unable to effectively penetrate or monitor provide safe haven within "human terrain." Most underground organizations rely upon—and seek to expand—this kind of safe haven. In addition, the covert and clandestine operational potential of cyberspace provides an environment which can be characterized as "virtual safe haven." Artificial safe havens can replicate actual restrictive terrain. However, artificial safe havens are only restrictive for as long as the risk of penetrating them remains unacceptable to the governing power's forces, or the government is simply unable to do so.

1-41. An important aspect of the civil considerations is the opportunity it presents for the resistance to access populations in enemy-controlled areas, to deliver information about the resistance and its objectives, and to establish beneficial lines of communications with key communicators. Active cultivation of relationships with key communicators can lower barriers and increase cooperation between U.S. forces and the resistance movement.

1-42. SOF units can sometimes negate the limitations of physical terrain to shape the operational environment. Planners should consider the full range of technological messaging platforms from social media sources (smart phone proliferation in third-world areas continues to grow) to Cold War legacy platforms consisting of shortwave transmitters to broadcast messages in areas where mountain ranges prevent line-of-sight frequency modulation radio broadcasting with messages targeted at a specific segment of the population. Shortwave transmitters (such as those used on EC-130J Commando Solo) coupled with

Chapter 1

Internet-based/cellular communications systems can broadcast messages to reach a greater percentage of the populace in the target area.

THE CRITICALITY OF THE FEASIBILITY ASSESSMENT

1-43. Planning remains limited until leadership validates certain assumptions. If operations proceed without a proper feasibility assessment, the likelihood of unintended consequences is high. To gain an accurate picture, operational personnel need to meet with indigenous personnel who represent the resistance forces. This meeting may take place inside the denied territory, in the United States, or in a third-party nation. Although meeting representatives in the United States or a third-party nation is safer for an assessment team, it also provides a less reliable assessment of potential capabilities and may be politically damaging if a covert or clandestine operation is exposed. Participation of all components is vital to enable an accurate assessment of potential resistance capabilities.

1-44. The assessment analyzes the feasibility, acceptability, and adequacy of a mission. This is an assessment based on mission, enemy, terrain and weather, troops and support available-time available and civil considerations (METT-TC) to determine if the necessary means and resources are available to meet mission requirements. It also addresses whether the potential gain or desired effect outweighs or otherwise justifies the potential losses or cost. Lastly, the assessment determines if achieving the desired objectives would accomplish the desired effects.

1-45. The normal areas of concern that make up a feasibility assessment are as follows:
- Are there groups that could develop into a viable force with assistance?
- Is the United States in contact with or can it make contact with individuals representing the resistance potential in an area?
- Are there any capable leaders whose goals are compatible with U.S. goals and who are willing to cooperate with the United States?
- Can the United States influence the leaders to remain compliant with U.S. goals?
- Are the groups' tactics and battlefield conduct acceptable by the standards established in FM 27-10 and to the U.S. and world population?
- Will the environment geographically and demographically support resistance operations?
- Is the enemy effectively in control of the population?
- Is the potential gain worth the potential risk? Is this group's participation politically acceptable to other regional partners?

1-46. All U.S. UW elements are able to assist the commander in answering these questions. They assist with individual perspectives for developing a particular resistance capability, as well as for an overall feasibility assessment.

1-47. Expatriates are a valuable resource, particularly in regions where the culture is largely unfamiliar or alien to a planner's frame of reference. However, planners should carefully ensure the individual's claims are valid. An expatriate's influence in a given country can be inversely proportional to the length of time he has been away from his former homeland. Although there are many reasons an expatriate might exaggerate his influence in a region and attempt to exploit the situation in his favor, he may be legitimately surprised to find his own assessment of his influence to be grossly inaccurate. During normal peacetime conditions, a person can spend years away from a country and expect to maintain their contacts and influence. This period significantly shrinks under the pressures of a harsh regime or occupying force.

1-48. While determining the feasibility of a potential campaign, planners must have clear objectives, a desired end state, and knowledge of exactly what level of support is available and acceptable. Without these specifics, negotiations with potential resistance forces are futile. If planners determine conditions are

unfavorable during the assessment, then they need to consider any measures that could transform the current situation into a more favorable one. For example, can the United States—
- Persuade a potential resistance group to cease unacceptable tactics or behavior?
- Persuade a coalition to accept a specific resistance group's participation under certain conditions?
- Degrade the enemy's control over the population?
- Bolster the will of the population to resist?
- Achieve desired objectives within the given time constraints?

1-49. SF can actively engage their resistance counterparts to encourage adherence to international norms of behavior and law. They can also change attitudes and beliefs about other groups participating in the resistance effort as part of unity and cohesion building. SF may leverage Psychological Operations (PSYOP) and Civil Affairs (CA) enablers to help engage and influence their partners. In circumstances where the resistance has sanctuary outside of denied territory, CA elements and MISO units may be able to link up directly to provide support.

1-50. Planners need to be careful of attempting to overcome a potential resistance shortcoming by creating resistance or guerrilla forces that are not indigenous. Historically, the United States has not had success creating and transplanting these types of resistance forces to an operational environment without an existing clandestine infrastructure that connects the local population to the foreign forces. (Appendix D discusses notable international practitioners of UW throughout history.)

This page intentionally left blank.

Chapter 2
Fundamentals

The guerrilla must move amongst the people as a fish swims in the sea.
Mao Tse-tung

Secret operations are essential in war; upon them the army relies to make its every move.
Sun Tzu

Separate definitions exist for resistance movements and insurgencies within the DOD and various academic communities. However, within this publication, the two terms convey a strategy of insurrection. Planners broadly use the term "insurgency" to describe the concept of achieving aims through a strategy of armed conflict and subversion against an indigenous government or occupying power. Planners use the term "resistance movement" to convey a unique type of insurgency that focuses on the removal of an occupying power. The difference in terminology is important to the concept of UW because planners must understand the significant differences in dealing with a resistance movement that forms in response to an occupying power, and an insurgency that grows over time out of discontent for a current regime. However, for purposes of this publication, insurgency, resistance, and revolutionary warfare are considered generally synonymous. In the context of this publication, insurgents are inherently indigenous. Consequently, planners should not categorize U.S. forces or foreign fighters supporting an insurgency or resistance movement as part of the insurgency. Instead, planners should categorize them as conducting UW as enablers, facilitators, advisors, or supporters.

WHY AND HOW POPULATIONS RESIST

2-1. Resistance generally begins with the desire of individuals to remove intolerable conditions imposed upon them by an unpopular regime or occupying power. Opposition toward the governing authority and for existing conditions that conflict with the individual's values, interests, aspirations, and way of life spread from the individual to the family, close friends, and neighbors. As a result, an entire community may possess an obsessive hatred for the established authority. Initially, this hatred may manifest as sporadic, spontaneous nonviolent and violent acts toward authority.

2-2. As the discontent grows, natural leaders—such as former military personnel, clergymen, college professors, local office holders, and neighborhood representatives—emerge to channel this discontent into organized resistance that promotes its growth. To maintain support for the cause, the population must continue to believe that the potential benefits are worth the risk of failure. Key to transitioning from growing discontent to insurrection is the perception by a significant portion of the population that they have nothing to lose by revolting and a belief that they can succeed. In addition, there must be a spark that triggers insurrection, such as a catalyzing event that ignites popular support against the government power and a dynamic insurgent leadership that is able to exploit the situation.

Chapter 2

CLANDESTINE RESISTANCE

2-3. People who outwardly follow their normal mode of existence conduct clandestine resistance. This type of resistance is organized, controlled, and conducts the following activities as groups and/or individuals:
- Political actions, propaganda, agitation, subversion, cadre generation, and organization-building.
- Espionage, intelligence-gathering, and infiltration of government and social organs.
- Sabotage and acts of politically driven targeted violence.
- Traffic in contraband, extortion, and other illicit fundraising activities.

2-4. The structure of an insurgency or resistance movement is similar to an iceberg (Figure 2-1, page 2-3). Most of the structure is below the surface and only the peak is visible. In building a resistance structure, insurgent leaders give principal attention to the development of a clandestine supporting infrastructure. As the clandestine infrastructure expands and becomes more effective, the entire effort will typically "surface" to include overt guerrilla actions. This predominantly covert and clandestine infrastructure works—
- Among the citizens' sociopolitical structures in rural villages, towns, and urban cities.
- Within the military, police, and administrative apparatus of government.
- Among labor groups; among educators, students, and other "intellectuals"; and within the press.

OVERT RESISTANCE

2-5. Individuals and groups who train along military lines perform overt resistance. Planners refer to this militant arm of the resistance movement as the guerrilla force. These elements make no secret of their existence or their objectives. However, resistance leaders compartmentalize the specific relationship of the guerrilla force to other components of the resistance movement to prevent compromise of the entire movement.

2-6. Each insurgency or resistance movement has its own unique characteristics based upon its strategic objectives, operational environment, and available resources. Insurgencies normally seek to change the existing social order and reallocate power within the country. Typical insurgent goals include—
- Removal of the established indigenous regime or occupying military power.
- Establishment of an autonomous national territory within the borders of a state.
- Extraction of political concessions that the movement cannot attain through nonviolent means.

Note: Many observers tend to equate "unconventional warfare" with "guerrilla warfare" and mistakenly exaggerate the importance of guerrilla warfare within the context of the entire UW campaign. In fact, guerrilla warfare is typically a relatively late-stage development in the course of a successful UW campaign. The political and organizational efforts of the underground are utterly indispensible to most successful instances of UW; they are the most difficult, most dangerous, and most time-consuming activities that constitute the overwhelming majority of a UW campaign. Guerrilla warfare is an important activity, but in the absence of effective political and organizational preparatory activities of the underground is almost certain to eventually fail.

It is of paramount importance that senior-level civilian decisionmakers, senior military commanders, JFCs, planners, and executing Soldiers, Sailors, Marines, Airmen, and interagency partners understand that when UW is not just an adjunct to anticipated major military involvement by the United States and its allies, that the function of guerrilla warfare is to support insurgent/resistance strategy. Guerrilla warfare and defeat of enemy forces on the battlefield is **not** the decisive event; the sum total result of political organization and mobilization is the decisive event in UW victory.

Fundamentals

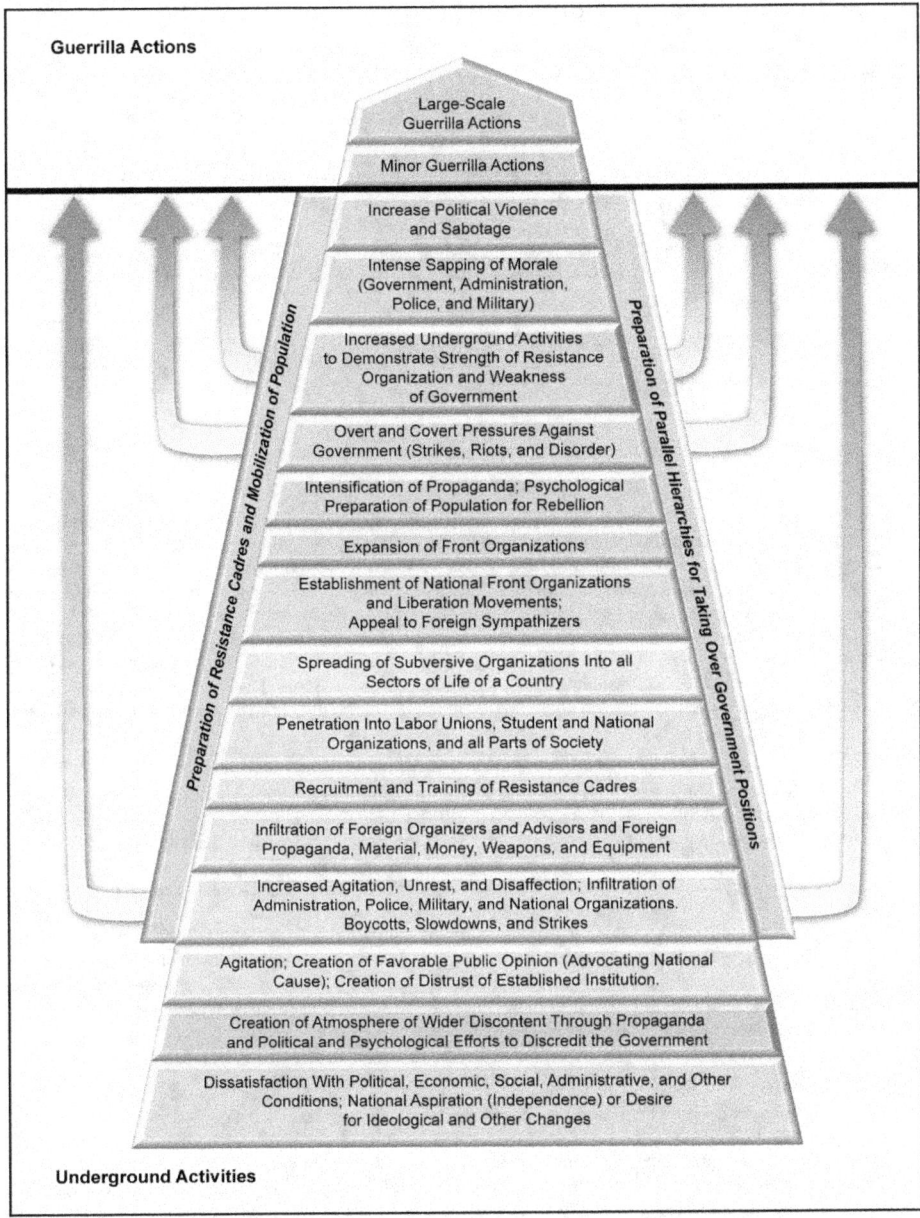

Figure 2-1. Structure of an insurgency or resistance movement

Dynamics of Successful Insurgencies

2-7. JP 3-24, *Counterinsurgency Operations*, lists eight dynamics of an insurgency. JP 3-24 includes internal support and organizational and operational patterns. FM 3-24, *Counterinsurgency*, lists six dynamics of an insurgency. FM 3-24 does not include organization and operational patterns. However, SF continue to use organization as one of the dynamics of an insurgency to understand the form, function, and logic of insurgent movements.

> *Note*: Appendix E is based on yet another set of evaluation criteria as used by a valuable UW resource—the *Casebook on Insurgency and Revolutionary Warfare, Volume 2: 1962–2009*. Although none of these lists are identical, each is valid and useful to understanding insurgency and resistance. The differences in interpretation are indicative of the variety of insurgent and resistance experience.

2-8. Seven dynamics are common to most successful insurgencies. These dynamics provide a framework for planners to analyze insurgencies. It is some combination of these dynamics that will generally transform popular disconnect into an organized and effective movement.

Leadership

2-9. A group committing random political violence is not an insurgency. In an insurgency, the group is committing directed and focused political violence. It requires strategic, operational, and tactical leadership to provide vision, direction, guidance, coordination, and organizational coherence. The insurgency leaders must make their cause known to the people and gain popular support. Their key tasks are to break the ties between the people and the government and to establish credibility for their movement. The leaders must replace the government's legitimacy with that of their own. Their ability to serve as a catalyst that motivates and inspires others to have faith in their cause is vital to the movement's growth. Their ability to organize and willingness to distribute power across the organization is vital to the long-term success of the movement. Organizations dependent upon key charismatic personalities to provide cohesion and motivation for the movement are vulnerable to disruptions if the enemy removes or co-opts those players.

Ideology

2-10. The insurgents must have a program that justifies its actions in relation to the movement's grievances and explains what is wrong with the status quo. The most important aspect of a successful insurgency is the viability of the message. It is essential that the message physically reaches the people and possesses meaning to their way of life. This makes the language, culture, and geography of the masses particularly important. Ideology is an important factor in unifying the many divergent interests and goals among the insurgency membership. As a common set of interrelated beliefs, values, and norms, ideology is used to manipulate and influence the behavior of individuals within the group. Ideology will serve as the rallying call for all members of the population to join the struggle. Major components of an ideology are commonly found in language, socio-economics, ethnicity, and religion. The ideology of the insurgency and the motivation of the insurgent must remain linked. Once delinked, the counterinsurgent will be able to address individual grievances and negate the unity of the insurgency.

Objectives

2-11. Although insurgency is a strategy, implementation requires intermediate objectives—specifically, strategic, operational, and tactical goals. At the tactical level, goals are most closely correlated with direct actions on the ground. These actions, in artfully designed combination, lead to operational goals. Insurgents need to carefully choose what efforts to undertake. For example, raids, ambushes, and supporting propaganda urging enemy forces to serve their tour quietly and go home alive, to consolidate into large bases away from the cities, or to stay to the roads for their own safety, achieve the goals of disrupting enemy control over territory and weakening enemy commitment to counterinsurgent strategy.

2-12. Operational objectives address how the insurgency will progress toward its strategic goal. Examples could include the following:
- Attaining a level of popular and passive support in a key region.
- Preventing government occupation or use of or transit through a "liberated" area.
- Gaining international recognition or external support.

2-13. The strategic objective is the desired end state. In general, the strategic objective is to gain concessions or remove the regime in power. Typically, the strategic objective is critical to cohesion among insurgent groups. It might be the movement's only clearly defined goal. Some examples of strategic goals are as follows:
- Anarchist—destructively eliminate government entirely.
- Egalitarian—impose universal equality.
- Traditionalist—resist change and return to a perceived norm.
- Pluralist—break a monopoly on political discourse.
- Secessionist—break off some section from the polity.
- Reformist—modify the application of laws and mores.
- Preservationist—safeguard valued institutions from change.
- Globalist—provoke supranational reorganization.
- Apocalyptic—act as catalyst for an envisioned end times.
- Utopian—impose a theoretical vision of man's perfection.
- Commercialist—facilitate greed through violent illegalities.

Environment and Geography

2-14. The environment and geography (including demographics) greatly affect an insurgency's strategies and tactics. Insurgencies may form their base in urban environments, rural environments, or a combination of both. By maintaining a combination of urban cells and rural bases, insurgencies can often take full advantage of the benefits of both models (urban and rural) without becoming constrained by the shortcomings of either model.

2-15. Insurgents located in rural areas enjoy the relative safety of remote terrain or safe havens, such as jungles or mountains. These geographical conditions make it possible for them to form larger guerrilla bands and conduct large-scale guerrilla operations. Disadvantages of a rural base are—
- Length and speed of communications and supply lines.
- Displacement of insurgents from the populace.
- Susceptibility of insurgents to conventional military counterguerrilla operations.

2-16. Urban insurgencies have overcome the lack of suitable restrictive terrain by operating within ethnic ghettos or enclaves within sympathetic densely populated urban areas. These areas often create safe havens that government forces are unwilling or unable to access. This type of urban basing requires a high degree of compartmentalization, which makes it more difficult for the group to train and organize for large-scale operations. Disadvantages of an urban base are—
- Greater susceptibility of infiltration and observation.
- Susceptibility of insurgents to police actions.

External Support

2-17. Historically, insurgencies do not succeed without some form of external support. Governments providing support to an insurgency normally share beneficial interests or common ideology with the

insurgency. Ethnic enclaves or diasporas in third-party countries can also provide significant support in terms of political voice, money, personnel, and sanctuary. This support can be in the form of—
- Moral or political support in the international forum.
- Resources, such as money, weapons, food, advisors, and training.
- Sanctuary, such as secure training sites, operational bases over a border, or protection from extradition.

2-18. With external support comes a degree of dependency on the foreign power. Insurgencies can view this as a disadvantage because the foreign power can then attempt to control or manipulate the insurgency to better serve its goals. To counterbalance the loss of the support from sympathetic foreign governments since the end of the Cold War, many groups have resorted to alliances with organized crime groups, narcotics trafficking, and kidnapping to raise funds. This tactic has proven extremely effective for generating revenue, but counterproductive to the original goals of the movements.

Phasing and Timing

2-19. Successful insurgencies pass through common phases of development. Not all insurgencies experience every phase, and progression through all phases is not a requirement for success. The same insurgent movement may be in different phases in separate regions of a country. Successful insurgencies can also revert to an earlier phase when under pressure, resuming development when favorable conditions return. A common failure of insurgencies and counterinsurgencies is the inability to adapt tactics when transitioning from one phase of a strategy to another.

2-20. The three-phase construct presented below is a historical representation of how insurgencies mature based on the Chinese communist model as articulated by Mao. From the perspective of the insurgent, Mao referred to the three phases as "Strategic Defensive," "Strategic Stalemate," and "Strategic Offensive." Admittedly, not all insurgencies and resistance movements are mass-based protracted conflicts. However, this classic insurgent model does provide an extremely useful analytical model to start from and a template for planners to communicate precisely an insurgency's stage of development.

Phase I—Latent or Incipient ("Strategic Defensive") Phase

2-21. During this phase, the leadership of the resistance develops the clandestine supporting infrastructure upon which all future efforts will rely. This is almost certain to be the most protracted phase and is by far the most important. The resistance organization uses a variety of subversive techniques to prepare the population psychologically to resist. Some techniques include propaganda, demonstrations, boycotts, and sabotage. Subversive activities include lending aid, comfort, and moral support to individuals, groups, or organizations that advocate the overthrow of incumbent governments by force and violence. All willful acts that are intended to be detrimental to the best interests of the government and that do not fall into the categories of treason, sabotage, or espionage will be placed in the category of subversive activity.

2-22. Subversive activities frequently occur in an organized pattern without any major outbreak of armed violence. Activities include the following:
- Recruit, organize, and train cadre.
- Infiltrate key government organizations and civilian groups.
- Establish cellular intelligence, operational, and support networks.
- Organize or develop cooperative relationships with legitimate political action groups, youth groups, trade unions, and other front organizations. This approach develops popular support for later political and military activities.
- Solicit and obtain funds.
- Develop sources for external support.

2-23. The goal is to prepare or transition the population into accepting insurgent direction and overt military operations (guerrilla warfare) as permissible, and to gain the support of the local population and weaken the power of the existing government. Operationally, the insurgents seek to gain large-scale popular support by multiple tactical efforts to convince local population organizations and neighborhoods to avoid collaboration with the government forces. This leads to a condition where the insurgency can expand operations without the risk of compromise by the local population. It is impossible for the insurgency to conduct the operations it desires without the population being aware of it. Success with all of these activities will allow the insurgency to expand into Phase II (Guerrilla Warfare).

Phase II—Guerrilla Warfare ("Strategic Stalemate") Phase

2-24. The objective of this phase is to degrade the government's security apparatus (the military and police elements of national power) to the point where the government is susceptible to defeat. A campaign of guerrilla attacks and sabotage degrade the government's military and police forces. Subversive activities continue to build and maintain support from the population. Pro-insurgency radio broadcasts, newspapers, and pamphlets openly challenge the control and legitimacy of established authority. Depending on perceived success and legitimacy, the resistance fighters or insurgents may achieve legal belligerent status from the international community if they meet the internationally accepted criteria.

2-25. In Phase II, unlike in Phase I, guerrillas need to gather forces, communicate and coordinate operations, conduct training, receive logistics, rest and hide after operations, and plan future operations. Their need for intelligence collection and security also increases in Phase II. As the guerrillas grow in numbers, so must the clandestine support mechanisms. Insurgent guerrilla forces are not yet strong enough to remain unhidden from government destruction. They attempt to achieve a rough parity between guerrilla forces versus security force combat power.

Phase III—War of Movement ("Strategic Offensive") Phase

2-26. The goal of the insurgency in Phase III is to bring about the collapse of the established government or withdrawal of an occupier after weakening it through the cumulative efforts of military and subversive actions. The insurgency does not necessarily need to transform into a conventional military, but it must position itself to defeat the government or occupying power. For example, the insurgency might degrade the enemy's capabilities to a point that an urban uprising against the presidential palace would topple the government. This tactic can only succeed if the insurgency effectively removes the military first.

2-27. As the insurgency gains control over the country, the insurgent leadership becomes responsible for the population, resources, and territory under its control. If the insurgency fails to plan and execute posthostility activities, the population may lose confidence in the insurgency and turn to the old government, a breakaway faction, or a splinter group of the insurgency.

2-28. Based on the conditions set earlier, an effective resistance or insurgency—
- Establishes an effective civil administration.
- Establishes an effective military organization.
- Provides balanced social and economic development.
- Mobilizes the population to support the resistance organization.
- Protects the population from hostile actions.

Failure to achieve these objectives may cause the resistance movement to revert to an earlier phase.

Organization and Operational Patterns

2-29. The organizational and operational pattern of a given movement is similar to its order of battle. From its outset, the organization has a concept of its development based on its goals. Although there are numerous traditional models for insurgencies (for example, conspiratorial, military-focused, urban, protracted popular war, or terror-based [not a methodology supported as a U.S. option]), the planner must avoid following a famous model without considering the way that model worked for its historical environment and if the model is appropriate for the current problem set. The structure of the organization largely depends on the available resources, security threat, and population distribution. All insurgencies are unique and rarely follow one model exclusively. It is unlikely the structure would resemble a uniform

Chapter 2

organization, such as the military, in which all units look relatively the same. Function takes precedence over form. Planners must understand the organization's—
- Various subordinate components and their orientation.
- Commands—down to the lowest tactical level.
- Supporting infrastructure.

2-30. The organization's most important component level is the local level, where it obtains and sustains support and manifests actions. Echelons above the local level coordinate all functions (political, military, external support, and so on). Overall command provides purpose and direction.

THE SEVEN PHASES OF UNCONVENTIONAL WARFARE

2-31. The seven phases of U.S.-sponsored insurgency are a conceptual template that planners use to aid understanding of a UW campaign effort. As a template, it merely serves as a guide for planning and execution. Not all phases will necessarily take place nor do they have to be executed in a rigid linear order. For example, U.S. forces may come to assist a large, previously established insurgent/resistance force with significant prior combat experience. In such a case, the organization, buildup, and employment phases may already be completed or ongoing before SOF ever infiltrate the JSOA. The seven phases are a model, not a checklist. With the exception of Military Information Support (MIS) forces, no other SOF will mirror the seven phases, but may support a specific portion or phase of the larger UW campaign. Regardless, operational personnel should understand how their efforts integrate with and contribute to the overall campaign plan.

PHASE I—PREPARATION

2-32. The preparation phase begins with the President and/or Secretary of Defense (SecDef) approval to execute a UW campaign. The primary purpose of this phase is to ensure that the insurgency, resistance, and the population are prepared to conduct and support a UW campaign. Years of preparation of the environment (PE) may have set the conditions for quick transition from shaping to preparation upon approval. This phase will include both IO targeting specific audiences to interagency efforts to determine goals, capabilities, and liabilities of the insurgency or resistance before making initial contact. If, during this phase, it is determined that the insurgency, resistance, or the population are not supportive of the effort, then the operation may be cancelled (for example, the proposed UW with Shia elements in Iraq in 2003).

2-33. The preparation phase must begin with a complete intelligence preparation of the operational area. UW operations need to include, but are not limited to, a thorough analysis of the resistance force's strengths, weaknesses, logistics concerns, level of training and experience, political or military agendas, factional relationships, and external political ties. Along with this data, any existing area studies of the target area are updated and refined. At a minimum, the area study includes governmental services, living conditions, and political, religious, economical, environmental, medical, and educational issues.

2-34. The USG begins to shape the target environment as far in advance as possible. The shaping effort may include operations to increase the legitimacy of U.S. operations and the resistance movement, building internal and external support for the movement, and/or setting conditions for the introduction of U.S. forces into the area of operations (AO). MIS assessments are particularly important during the preparation phase because they provide planners and executing units with vital information on possible insurgent leaders and key communicators that have psychological relevance with the population. Operatives can conduct these activities proactively (in areas under the control of adversarial regimes) or reactively (immediately following an act of aggression against an ally's territory, such as an invasion). The population of a recently occupied country may already be psychologically ready to accept U.S. sponsorship, particularly if the country was a U.S. ally before its occupation. In other cases, psychological preparation may require a protracted period before yielding any favorable results.

2-35. CA forces conduct in-depth area studies and analysis addressing the full extent of the UW operational area across the political, military, economic, social, information, and infrastructure (PMESII) variables. This analysis includes identification of key civil infrastructure, civil authorities, and general conditions likely to affect the population's conduct during the execution of a UW campaign, such as government services; living conditions; and political, religious, economical, environmental, medical, and educational issues. CA elements begin executing civil information management in order to fuse historical and newly obtained civil information with intelligence to develop a holistic understanding of the operational environment. This process ultimately serves to identify key civil vulnerabilities that can be exploited or otherwise addressed in later phases.

PHASE II—INITIAL CONTACT

2-36. Before the USG decides to render support to a resistance, it establishes contact with representatives of a resistance organization to assess the compatibility of U.S. and resistance interests and objectives. This assessment is largely a political negotiation between the USG and the resistance organization. Once the USG establishes compatibility, it assesses the resistance potential. During the initial contact, planners may arrange for the reception of a pilot team. (A *pilot team* is **a deliberately structured composite organization comprised of Special Forces operational detachment members, with likely augmentation by interagency or other skilled personnel, designed to infiltrate a designated area to conduct sensitive preparation of the environment activities and assess the potential to conduct unconventional warfare in support of U.S. objectives.**) If possible, planners may exfiltrate a resistance representative, referred to as an asset, from the operational area to brief the pilot team during its planning phase and possibly to accompany the team during their infiltration into the operational area and linkup with resistance forces.

2-37. The pilot team conducts detailed area assessments to expand its understanding of the operational environment, particularly human terrain analysis. This analysis provides information on the degree of support for the UW effort among the local populations. MISO staff planners provide the pilot team with area- and population-specific information requests to facilitate the evaluation of indigenous information capabilities and the determination of the level of support necessary to fully develop those capabilities and increase their operational effectiveness.

2-38. CA planners supporting a pilot team refine previous area studies in order to further identify factors affecting civil infrastructure and ultimately the civil center of gravity in the UW operational area. CA personnel also make initial liaison with civil officials within or outside the UW operational area, conduct civil reconnaissance and assessments of critical infrastructure. CA personnel tie into existing persistent engagement programs for access to civil information management data, personal and professional relationships, and general area knowledge to take advantage of existing civil networks, thereby allowing for expanded ability to understand and favorably influence the relevant population and its actions.

PHASE III—INFILTRATION

2-39. During the infiltration phase, a pilot team clandestinely infiltrates the operational area in order to link up with the resistance force and conduct or confirm a feasibility assessment. If the operational assessment is favorable, the pilot team coordinates for the infiltration and reception of follow-on SF teams and supplies. MISO forces, if attached to follow-on SF teams, can provide operational detachments with an early information capability while developing an indigenous capability. As follow-on teams infiltrate the operational area and link up with their respective resistance-force counterparts, they begin their own operational assessment to confirm or deny the assumptions of the overarching UW campaign plan.

2-40. Depending on the situation, circumstances may dictate the infiltration of operational detachments without the benefit of a prior pilot-team effort, a trusted asset, or a completed feasibility assessment. In this case, the operational detachments will perform many of the required functions normally accomplished by the pilot team. If this occurs, the operational detachments may need to adjust their infiltration plans to account for the higher risk of infiltration without the benefit of a reception coordinated by other U.S. personnel.

2-41. Successful infiltration of operational detachments and tailored command elements to ensure countrywide or theater-wide unity of effort and apportionment of resources represents a decisive point for the operation, as well as a period of increased operational risk. Infiltration plans and tactics need to remain focused on successfully achieving this decisive point. Once the operational detachments successfully link up with resistance forces, they must conduct the same level of operational assessment that the pilot team normally conducts before the implementation of their plan to render support.

2-42. CA personnel attached to SF teams conduct civil reconnaissance, link up with counterparts, verify accuracy of historical or previously obtained civil information, provide the detachment with relevant civil information across the PMESII variables, and coordinate civic-action programs in close collaboration with the interagency, intergovernmental organizations (IGOs), and nongovernmental organizations (NGOs), where appropriate. Programs are designed to minimize civilian interference with guerrilla forces and the shadow government, while enhancing their public image and stability. CA can gain and maintain access to denied areas through rapid and low-cost civil initiatives that directly benefit the targeted population. At the same time, CA teams identify and assess infrastructure critical to regime control over the population. Interdiction by guerrilla forces can be augmented considerably as a result, further separating the population from regime control.

2-43. In limited-war scenarios where the infiltration of U.S. personnel is undesirable, planners could exfiltrate indigenous resistance personnel out of the target area, provide training in specific required skills, and infiltrate the personnel back into the target area to function as cadre capable of conducting operations or training other resistance forces. The enemy's level of control over the population and the environment affects how long planners can keep resistance personnel away from their region before they begin to disconnect from the local population.

PHASE IV—ORGANIZATION

2-44. Once U.S. advisors link up with resistance leadership, the objective is to determine and agree upon a plan to organize the resistance for expanded operations. In addition to physical preparations, this entails a confirmation of mutual objectives and prior agreements. This requires a period of rapport-building to develop trust and confidence, as well as a period of discussion of expectations from both sides.

2-45. Before a resistance organization can successfully engage in combat operations, its leadership must organize an infrastructure that can sustain itself in combat and withstand the anticipated hostile reaction to armed resistance. During the organization phase, the resistance leadership develops a resistance cadre to serve as the organizational nucleus during the buildup phase. The operational detachments assist the resistance leadership in this endeavor by providing training and advice to prepare for the eventual buildup of the resistance organization.

2-46. MISO forces assist in this phase by organizing and developing a new (or building upon an existing) resistance information capability. Once the resistance develops this capability, it can incorporate information elements into insurgent and auxiliary organizations. Development includes training and mentoring cadre and advising on their employment. The ultimate goal is the employment of resistance information capabilities that are increasingly self-sustaining.

2-47. CA build trust with and provide their guerrilla counterparts training and equipment necessary for the successful employment of CA and civil-military operations (CMO) assets among the targeted population. Instruction in and provision of the knowledge and tools necessary to alleviate the population's suffering is critical to UW success. Delivery of a higher standard of living than that provided by the regime will have far-reaching psychological and operational advantages for the shadow government and guerrilla forces. Attached CA elements will provide the detachment with a detailed assessment of resistance governance and civil administration capacity.

2-48. Many guerrilla leaders may not enthusiastically accept U.S. advisors but may tolerate them as a precondition for other U.S. support, such as logistical aid. They may harbor suspicions that the U.S. intentions are purely self-serving or lack the resolve to maintain their commitment long-term. Guerrilla leaders are cautious of placing too much reliance on U.S. promises. It is the challenge of the operational detachments' leadership to gain the confidence of the resistance leadership and demonstrate the value of cooperation toward their mutual goals. The operational detachments explain their capabilities and limitations and begin to assist the resistance leadership with the development of the organization. Although rapport eases every aspect of operations, operational personnel must not perceive rapport as the goal but rather a means to a goal. The goal is a strong relationship in which the SOF leaders are trusted advisors who can influence the direction of events. The amount of influence an advisor attains is directly proportional to the total sum of three factors between himself and his counterpart. These factors equate to rapport, credibility, and continued belief in the value of the relationship.

2-49. The resistance leader and U.S. advisors must agree upon command arrangements within the bounds of higher-level political and military agreements. The specifics of a resistance organization depend on local conditions. UW requires centralized direction and decentralized execution under conditions that place great demands on the resistance organization and its leadership. Armed rebellion inherently creates an ambiguous and unstructured environment. No two resistance organizations need the same degree or level of organization. All levels of command should consider the following factors when advising the resistance leadership concerning organization:

- The effectiveness of the existing resistance organization.
- The extent of cooperation between the resistance organization and the local population.
- The level of hostile activity and security measures.
- The political boundaries, natural terrain features, potential targets, and population density of the operational environment.
- The religious, ethnic, political, and ideological differences among elements of the population and competing resistance organizations.
- The proposed type and scope of combat operations.
- The degree of U.S. influence with the resistance organization.

2-50. It is important that the operational detachments understand and report the strengths and weaknesses of the resistance group. With detachment feedback, higher commanders can develop the UW campaign plan to effectively leverage the strengths of the different groups while mitigating the inherent weaknesses. MIS and CA assessments can augment SF evaluations and aid in understanding these strengths and weaknesses. These assessments aid in understanding the cultural, religious, economic, and social factors affecting the operational environment and the resistance movement. This analysis also provides key insights into relationships and other influences affecting the behavior of targeted groups.

2-51. Planners cannot automatically fix the organization of the guerrilla force according to conventional tables of organization and equipment. Guerrilla force missions and tactics dictate a simple, mobile, and flexible organization capable of rapid dispersion and consolidation. Each unit must function as an independent organization with its own intelligence, communications, and logistical systems. Guerrilla organization normally determines auxiliary organization. Planners should compartment all auxiliary functions from one another and from the guerrilla force that the auxiliary supports. MIS elements develop complementary guerrilla and auxiliary information elements. This enables them to provide training in tactics, techniques, and procedures that match the capabilities of those organizations.

2-52. Once U.S. advisors have an accurate assessment of the situation on the ground, a good working relationship, and a concept for expanded operations, resistance leaders and planners work out the specifics of the concept. The two parties work out these specifics at all levels, clearing them through U.S. interagency channels to ensure concept agreement.

PHASE V—BUILDUP

2-53. During the buildup phase, the resistance cadre improves the organization's clandestine supporting infrastructure in order to prepare for expanded offensive operations. When the organization begins to conduct operations of a wider scope and across a wider area, many of these operations will draw attention from counterguerrilla forces. The organization must have the supporting clandestine infrastructure to prepare for and sustain these operations. Possible expansion efforts to prepare the area complex for future combat operations include—

- Increasing intelligence, counterintelligence, and early warning networks.
- Expanding recruiting efforts without compromising security of the force or operations.

- Developing and expanding underground capabilities to conduct information, deception, and sabotage activities.
- Expanding the supply distribution and cache networks.
- Expanding guerrilla training programs for new recruits or on new materiel supplied by the operational detachments, such as mortars; antitank systems; machine guns; demolitions; medical gear; chemical, biological, radiological, nuclear, and high-yield explosives protective gear; and so on.
- Developing mechanisms and networks to support evasion and recovery of coalition forces, such as downed airmen.

2-54. During this phase, the resistance force may conduct limited offensive operations to gain confidence and experience, to procure needed materiel and supplies, or to confuse and harass the enemy forces in their area. However, the emphasis remains on developing the resistance infrastructure to support future operations. The resistance leadership must not let the organization expand beyond its organic capability to sustain itself.

2-55. All parties must carefully consider the type and scope of future combat operations that are likely to achieve each organization's desired plans, as well as the plans of the overarching resistance command element in conjunction with or supporting U.S. efforts. All levels of command must be aware of the theater commander's intent for resistance combat operations. During isolation planning, units will focus on some likely strategic, operational, and tactical targets. It is unlikely the unit would have had the fidelity to determine appropriate objectives and targets that achieve the theater commander's desired effects before linkup with the resistance leadership. After advisors and the resistance leadership agree upon a support plan for the overall UW campaign, they must determine which actual targets achieve the desired goals.

2-56. The first step toward determining appropriate targets, objectives, and other supporting efforts is to establish an understanding of the enemy's capabilities and intentions:
- What are the enemy's tactical, operational, and strategic capabilities and efforts to ensure control over the population?
- What are the enemy's centers of gravity?
- What are the enemy's psychological vulnerabilities?
- Where is the enemy vulnerable to guerrilla or underground operations?

The enemy situation may change drastically from the current time to the time of a coalition invasion.

2-57. Depending on the length of the campaign effort, it is highly probable that conditions on the ground have changed since SOF assets infiltrated the JSOA. The intelligence derived from the resistance in this phase is critical to updating theater planning efforts.

2-58. Each level of headquarters (HQ) must identify potential objectives or lines of operations with their resistance counterparts as part of the overall UW campaign plan. Units ensure the resistance efforts complement the theater campaign plans. Planners must take care to minimize the communication signature during coordination so that they do not burden or jeopardize resistance units while satisfying conventional planning procedures.

2-59. Once planners determine the type and scope of operations, they develop supporting capabilities specific to those efforts. Tasks outlined in the UW campaign plan or operation order drive supporting capabilities. The operational detachments may start to coordinate for specific supplies via airdrop or other resupply methods. Planners need to prioritize resupply efforts for materials that forces cannot procure by other means. Every resupply operation comes with the risk of exposure and potential loss of the supporting apparatus. Planners need to develop and emplace capabilities without compromising the organization or future operations. If resistance efforts are to support a pending coalition D-day (start of the coalition invasion), the capabilities—whether guerrilla, auxiliary, or underground—need to include notification and activation procedures that allow the synchronization of efforts with conventional forces.

2-60. CA teams advise guerrilla forces throughout the course of expanded operations, mentoring them to implement humanitarian assistance, disaster relief, and populace and resources control designed to address civil vulnerabilities and gain the favor of a supporting population IAW the UW campaign plan. Critical to this phase are definitive assessments of civil centers of gravity and critical infrastructure, as well as engagement with key civic leaders and decisionmakers within the shadow government and the supporting population. During Phase V, a civil-military operations center (CMOC) may be established to direct and coordinate humanitarian efforts in support of U.S. goals. The CMOC must be in a secure area to be effective.

2-61. In restrictive environments, CA forces may accomplish their mission entirely from outside a country through proxy elements. If U.S. involvement is clandestine, CA personnel can operate a CMOC in an extra-national location, coordinating displaced person/refugee camp operations and facilitating vetted third-party elements (for example, medical, humanitarian, or construction IGOs or NGOs) neutral or sympathetic to the resistance. If the United States is overtly supporting the resistance, the CMOC will collocate with U.S. forces and be able to establish and enhance governmental and administrative mechanisms critical through Phases VI and VII.

PHASE VI—EMPLOYMENT

2-62. During the employment phase, the resistance force initiates an expanded scope of offensive operations to achieve the desired effects. The desired effects can range from causing the enemy to commit limited resources away from a pending invasion area, support to a pending invasion area (when there is large-scale conventional force commitment), or in the case of a separate insurgency (when there is no large-scale conventional force commitment), the achievement of specific strategic politico-military objectives. The main activities in this phase consist of interdiction and MISO.

2-63. Planners need to consider the types of targets guerrilla forces attack. Operations may include the following:

- Interdicting lines of communication.
- Attacking enemy command and communications nodes.
- Targeting rear-area infrastructure and high-payoff targets, such as munitions and fuel depots; rail yards; airfields; waterways; power plants; and radio, television, and other mass media facilities.
- Seizing key objectives—such as choke points, prisoner or concentration camps, and critical national infrastructure—for very limited periods of time.

2-64. MISO seek to achieve two goals simultaneously in this phase: exploit guerrilla successes for their maximum psychological effect to erode enemy morale and decrease internal and external support. This exploitation can increase the morale of resistance forces and auxiliaries, which can further increase their operational effectiveness. In addition, success brings positive attention to the movement and increases support from the indigenous population and external supporters in terms of logistics, intelligence, and recruitment. This exploitation may also increase dissension, desertion, and surrender of enemy forces, further decreasing their operational effectiveness. MISO can further exploit enemy reprisals against populations or the guerrillas to separate the population from the enemy government or occupying forces.

2-65. Civil Affairs operations (CAO) during the employment phase works toward a favorable outcome following interdiction by managing impact and consequences of actions against regime forces and infrastructure, as well as mitigating guerrilla actions against neutral or hostile noncombatants. Continual engagement of civil figures and organizations, as well as interagency collaboration with nongovernmental organizations and humanitarian government agencies potentially operating in the battlespace, are also critical components of the ongoing effort to win over the populace to the shadow government's side. CAO executed during this phase are directed by the CMOC, and focus on surging foreign humanitarian assistance; populace and resources control, to include dislocated civilian operations; and deconflicting intergovernmental/nongovernmental organizations' operations to provide unity of effort.

2-66. Physical attacks that resistance forces conduct can significantly alter psychological effects during this phase. Planning actions for psychological effect is a deliberate process requiring thorough analysis, detailed coordination, and careful execution. Although this process is time- and labor-intensive, the effects can potentially shape the course of the entire UW operation in a profound manner. SF units on the ground

coordinate and synchronize these efforts to ensure a complementary effect, first with partner force and then conventional force efforts. Other supporting efforts may include—
- Gathering and reporting vital intelligence to coalition forces.
- Assisting in the evasion and recovery of isolated personnel (downed aircrews).
- Reconnoitering and receiving airborne, air assault, or amphibious invasion forces.

2-67. MISO can enhance the effects of these supporting efforts. They can increase actionable intelligence obtained from key segments of the population through persuasive messages that increase sympathy and support for the resistance movement. Information on rewards and other messages can persuade target groups to aid the evasion and recovery of isolated personnel. In addition, MISO can assist in building local networks that provide support for incoming invasion forces by consistently emphasizing the benefits of supporting the UW effort and highlighting the negative aspects of the enemy government or occupying power.

2-68. In a scenario involving a pending conventional force invasion, planners synchronize combat operations around an undisclosed D-day. For operations security, planners may not tell U.S. personnel or resistance forces the specific time and date of D-day, which would make synchronization particularly critical and challenging. Instead, SOF elements and resistance forces in the JSOA may be given coded signals and contact schedules to monitor for instructions to initiate precoordinated operations in support of the conventional invasion. Planners may choose a defining date or event, such as the anniversary of a civil disturbance, to motivate the guerrilla force and increase popular support for operations. Careful selection of the D-day can have significant psychological effects on UW operations. The HQ coordinates notification procedures and the amount of advance warning they require with theater HQ before the infiltration of operational detachments. Resistance forces require time to notify and assemble their forces, recover equipment, and move into position. Resistance forces usually must accomplish these tasks without the benefit of secure technical communications gear and without alerting enemy forces. Based on the expectation of liberation by invasion forces, the resistance can assume the risk associated with initiating more offensive operations than is normally permissible. When properly coordinated, these offensive operations can have a devastating disruptive effect on an adversary's combat capability. However, if the resistance initiates its efforts too early, they will alert enemy forces and possibly initiate retaliation. If the resistance initiates its efforts too late, they may not have the required effect to be of value.

2-69. In a limited-war scenario, this phase still consists of a campaign of guerrilla warfare and subversion, but forces execute them in a slightly different manner. Combat operations generally do not focus around a single culminating D-day event. Forces conduct these operations over a protracted period of time, with the intent of slowly eroding enemy strength and morale. Guerrilla attacks and acts of sabotage and subversion drain the hostile power's morale and resources, disrupt its administration, and maintain the civilian population's morale and will to resist. By repeatedly attacking multiple and widely dispersed targets, the resistance organization confuses, frustrates, and demoralizes hostile forces. Such attacks force the hostile power to divide its reaction and reinforcement capabilities. This slowly creates an increasing demand on the enemy to spend a disproportionate amount of strength to maintain its existing state of control over the population. In either a general-war or limited-war scenario, advisors ensure that resistance activities continue to support the objectives of the U.S. unified commander, mindful that resistance objectives are rarely identical to those of the United States.

2-70. Some planners assume the goal is to enable resistance forces to transform and equate to additional conventional infantry units. Resistance forces are not meant to morph into copies of U.S. or allied conventional forces. Campaign planners must understand the capabilities and limitations of the resistance forces, as well as the associated advantages and disadvantages. As conventional forces near areas with resistance forces, the joint task force (JTF) coordinates linkup between the two forces. Planners arrange how linkup will occur before infiltration of the Special Forces operational detachment A (SFODA). Successful linkup is critical to avoid friendly fire incidents and to give the maneuver commander the benefit of the resistance force's intimate knowledge of the local environment. This normally requires the placement of an element with the advancing conventional force HQ. After linkup, resistance forces revert to some variety of coalition or national control. These forces may demobilize or transition to a regular military or security force.

Phase VII—Transition

2-71. Elements conduct UW until they have met the mission objectives. In some cases, these objectives will include removing the hostile government or occupying power entirely. In such cases, U.S. support to the new government transitions from UW to foreign internal defense. At this point, it is critical to shift mindsets from defeating the adversary regime to protecting the newly installed government and its security personnel from insurgency, lawlessness, and subversion by former regime elements that attempt to organize resistance. Planners should have addressed transition planning in the feasibility assessment that formed the basis of the determination to support the resistance organization. Elements must honor their commitment to promises made before transition. Resistance forces are more likely to demobilize or transition if—

- The postconflict government reflects their original expectations.
- The belief exists that elements will honor promises made before transition, specifically in terms of benefits, such as back pay for service to their country or future employment.

2-72. During the period of conflict, it is possible that attitudes of a given force change or allegiances collapse. It is common for isolated guerrilla units to lose sight of their original objectives in order to maintain a lifestyle that is now more familiar than their preconflict lives. For this reason, all levels need to monitor attitudes and reinforce the end state objectives throughout the course of the campaign. MISO units are particularly useful in disseminating information about the benefits of demobilization, reintegration, and a reestablishment of society and civilian life. The manner in which the transition occurs affects the postwar attitudes of the people and the government toward the United States. Perhaps the greatest danger in transition is the possibility that former resistance members may resort to factional disputes, banditry, or subversion of the new government. The new government must make every effort to reorient and absorb former resistance members into a peaceful society and gain their acceptance. To achieve this goal, the new government must—

- Bring arms and ammunition under government control.
- Assist resistance members in returning to civilian life.
- Use resistance members as local militias or the base for future police and army forces.
- Take positive measures to prevent resistance members from beginning or participating in further political upheaval.

2-73. Because of their knowledge of resistance organization and history, SF teams initially remain in their operational areas to assist in the demobilization effort or in the transition of former resistance forces into national regular forces. SF personnel may serve as trainers and advisors to newly formed counterguerrilla or counterterrorist units, particularly if former resistance forces reject transition and continue to violently oppose the new indigenous government. During transition and demobilization, CA units are critical assets in helping the new government meet the needs of the former resistance forces and their families. MISO are essential during this phase in explaining the demobilization process and promoting loyalty of guerrilla forces to the new government as part of continuing efforts to maintain support for the movement's transition to a working government. The key to long-term strategic success in UW is the planning and execution of SOF postconflict responsibilities.

2-74. Finally, it should be noted that this transition phase applies to a UW campaign that proceeds to the complete overthrow of a government or the ousting of an occupying power. Where the objectives of the UW campaign are much more limited and only seek to coerce or disrupt adversary government behavior, there may not be a transition phase as described above at all. In such cases, transition may consist of the relationship with indigenous elements being terminated, going dormant, or being handed off to other USG agencies or foreign partners.

2-75. In Phase VII, CA elements focus their governance and development efforts on assisting the new government to meet the needs of its constituency, including repatriated resistance forces. As the regime is replaced or reformed into a new U.S.-backed government, it is critical for CA elements to continue to operate closely with their host-nation partners in order to redress civil grievances and reconstitute civil infrastructure and organizations. Demobilization of mass mobilization efforts is fundamental to successful transition operations. CA will focus operations on smaller, operationally critical population centers hard-hit by conflict and vital to survival of the new government. This normally entails expanded collaboration with aid-based organizations, such as the United States Agency for International Development (USAID), as well as the new host-nation government.

THE COMPONENTS OF AN INSURGENCY

2-76. There are three primary components of an insurgency: the underground, auxiliary, and the guerrillas. The goals and objectives of the insurgency will determine the level of development and relationships among the underground, auxiliary, and the guerrillas. The underground is always present and is the first component of the insurgency to form.

THE UNDERGROUND

2-77. The underground is a cellular organization within the resistance movement or insurgency that has the ability to conduct operations in areas that are inaccessible to guerrillas, such as urban areas under the control of the local security forces. The underground can function in these areas because it operates in a clandestine manner, which prevents it from receiving legal belligerent status under any international conventions. Examples of underground functions include the following:
- Intelligence and counterintelligence networks.
- Subversive radio stations.
- Propaganda networks that control newspaper or leaflet shops, Web pages, and social media sites, blogs, and postings.
- Special materiel fabrication, such as false identification, explosives, weapons, and munitions.
- Control of networks for moving personnel and logistics.
- Individuals or groups that conduct acts of sabotage in urban centers.
- Clandestine medical facilities.

2-78. Underground members normally are active members of the community, and their service is a product of their normal life or position within the community. They operate by maintaining compartmentalization and delegating most risk to their auxiliary workers. The functions of the underground largely enable the resistance movement to affect the urban areas.

2-79. The operations cell is usually composed of a leader and a few cell members operating directly as a unit (Figure 2-2, page 2-17). A cell can be as large as required to conduct a specific mission, but is generally kept as small as possible for security reasons. The responsibilities of each cell member will be as defined by the cell leader and the cell's assigned functions. However, cell member responsibilities are also shaped by the cell member's societal status and skill sets. Operational cell members may or may not know each other or work together. Wherever possible, organizational security is enhanced by structuring intermediaries between operational members and higher-echelon leaders. This system of intermediaries permeates the entire insurgent structure all the way to the organization's strategic leadership.

2-80. The intelligence cell exhibits similar organizational and security characteristics as operational cells. However, the need for cell security is of critical importance; the intelligence cell leader is seldom in direct contact with the members of the cell, and the members are rarely in contact with one another (Figure 2-3, page 2-17). Methods of assigning missions down and reporting information up these structures maximize third-party intermediaries or mechanical communication methods which avoid human-to-human direct contact altogether.

Fundamentals

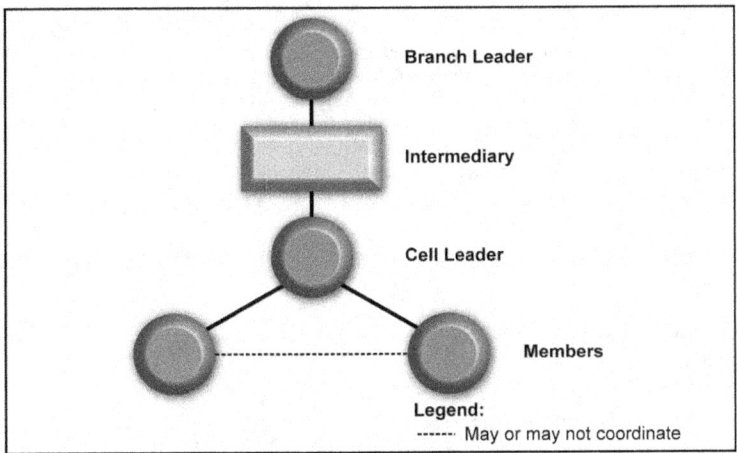

Figure 2-2. The operations cell

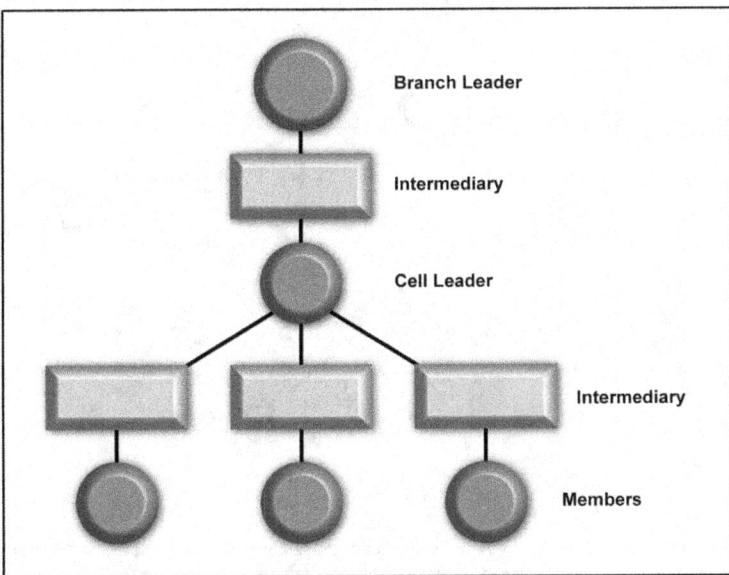

Figure 2-3. The intelligence cell

THE AUXILIARY

2-81. The auxiliary refers to that portion of the population that provides active clandestine support to the guerrilla force or the underground. Members of the auxiliary are part-time volunteers that have value because of their normal position in the community. Soldiers should not think of the auxiliary as a separate

6 September 2013 ATP 3-05.1 2-17

Chapter 2

organization but as a different type of individual providing specific functions as a component within an urban underground network or guerrilla force's network. These functions can take the form of logistics, labor, or intelligence collection. Auxiliary members may not know any more than how to perform their specific function or service that supports the network or component of the organization. In many ways, auxiliary personnel assume the greatest risk. They are also the most expendable element within the insurgency. Insurgent leaders sometimes use auxiliary functions to test a recruit's loyalty before exposing him to other parts of the organization. Auxiliary functions are like embryonic fluid that forms a protective layer, keeping the underground and guerrilla force alive. Specific functions include the following:

- Logistics procurement and distribution (all classes of supply).
- Labor for special materiel fabrication.
- Security and early warning for underground facilities and guerrilla bases.
- Intelligence collection.
- Recruitment.
- Communications network staff, such as couriers and messengers.
- Propaganda distribution.
- Safe house management.
- Logistics and personnel transport.

2-82. The auxiliary cell is commonly found in front groups or in sympathizers' organizations. It contains an underground cell leader, assistant cell leaders, and members (Figure 2-4).

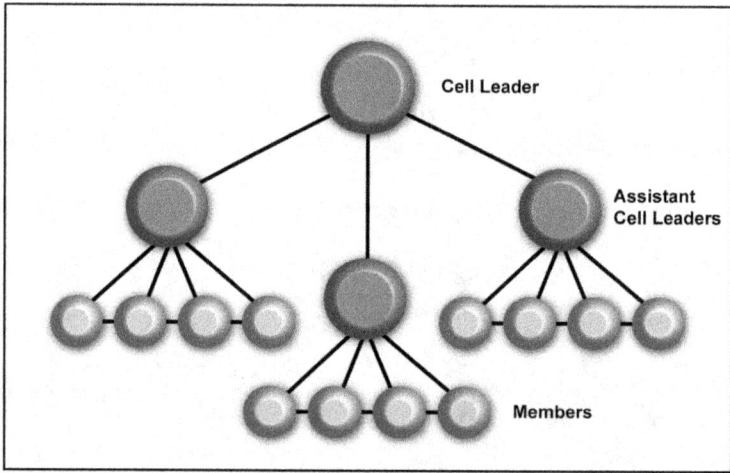

Figure 2-4. The auxiliary cell

2-83. The key idea of parallel cells is that many activities can be conducted that support a goal known only to a higher controlling echelon leader. For example, an underground planning to conduct an attack on a government facility may assign more than one intelligence cell to surveil the intended target. Although both cells have the same assignment, each cell is unaware of the membership, presence, or mission of the other cell. Using parallel cells provides the overall organization operational redundancy. Although unknown to the executing cells, redundant secondary parallel cells are frequently set up to support a primary cell (Figure 2-5, page 2-19).

Fundamentals

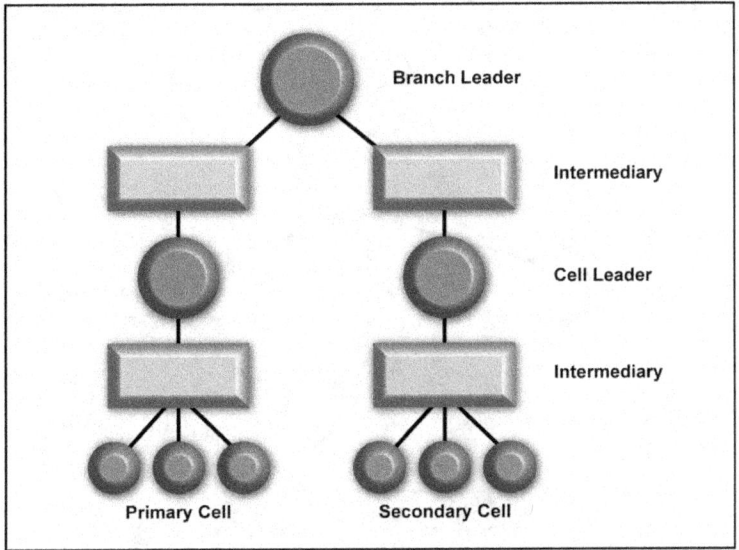

Figure 2-5. Parallel cells

2-84. The cells in series provide a division of labor in order to carry out underground and auxiliary functions, such as the manufacture of weapons, supply, escape and evasion, propaganda, and printing of newspapers. The task assigned to a particular cell must transition or carry over (depicted by arrows) to the next cell in order to accomplish the function in its entirety. For example, cell 1 purchases certain items, cell 2 assembles the items, and cell 3 distributes the assembled item (Figure 2-6, page 2-20). Based on the assigned mission, cell members do not communicate directly with one another. However, cell leaders will communicate indirectly through intermediaries.

THE GUERRILLAS

2-85. Guerrillas are the overt military component of a resistance movement or insurgency. As the individuals that engage the enemy in combat operations, guerrillas typically have a significant disadvantage in terms of training, equipment, and firepower. For all their disadvantages, guerrillas have one advantage that can offset this unfavorable balance—the initiative. In all his endeavors, the guerrilla commander must strive to maintain and protect this advantage. The guerrilla only attacks the enemy when he can generate a relative, if temporary, state of superiority. The guerrilla commander must avoid decisive engagements, thereby denying the enemy the opportunity to recover, regain their actual superiority, and use it against the guerrilla force. The guerrilla force is only able to generate and maintain the initiative advantage in areas where they have significant familiarity with the terrain and a connection with the local population that allows them to harness clandestine support.

2-86. Depending on the degree of control over the local environment, the size of guerrilla elements can range anywhere from squad to brigade-size groups or larger. In the early stages of an insurgency, the guerrilla force's offensive capability might be limited to small standoff attacks. As the guerrilla force's base of support from the population grows, its ability to challenge government security forces more openly with larger-scale attacks increases. At some point in an insurgency or resistance movement, the guerrillas may achieve a degree of parity with host-nation forces in certain areas. In these cases, units may start openly fighting, rather than as guerrilla bands. In well-developed insurgencies, formerly isolated pockets of resistance activity may eventually connect and create liberated territory, possibly even linking with a friendly or sympathetic border state.

Chapter 2

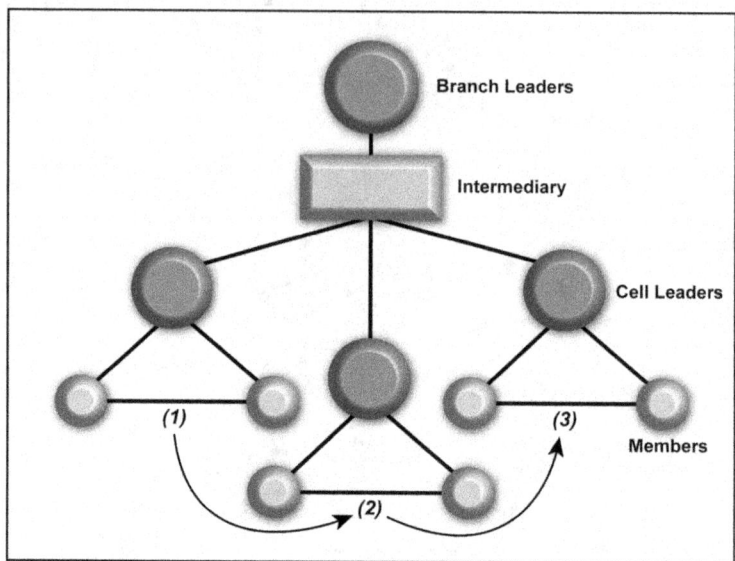

Figure 2-6. Cells in series

2-87. It is important to use the term "guerrilla" accurately in order to distinguish between other types of irregular forces that might appear similar but are in fact something entirely different, such as militias, mercenaries, or criminal gangs. A *guerrilla* is an **irregular, predominantly indigenous member of a guerrilla force organized similar to military concepts and structure in order to conduct military and paramilitary operations in enemy-held, hostile, or denied territory.** Although a guerrilla and guerrilla forces can exist independent of an insurgency, guerrillas normally operate in covert and overt resistance operations of an insurgency.

ADDITIONAL COMPONENTS OF AN INSURGENCY

2-88. Additional components and elements may be found in an insurgency. Some typical additional elements are discussed in the following paragraphs.

LEADERSHIP AND COMMAND

2-89. Leadership is not a separate type of component as much as it is a function. Strategic leadership, which conceptualizes the ends, ways, and means of the insurgent or resistance strategy, can be located anywhere—in the AO or in another country—so long as they have a means of conveying strategic guidance to insurgent elements in the AO. Some relatively small number of leaders will also likely be outside of the AO functioning as part of one or more insurgent support networks.

2-90. Most of the organizational leadership, however, will of necessity be located inside the AO, and often immersed in and influencing the population. Operational and tactical leadership is essential to the creating, buildup, and employment of all underground, auxiliary, and guerrilla elements. For this reason, most successful insurgencies and resistance movements devote much time and resources into developing trained, indoctrinated, and disciplined cadre who can magnify unifying themes, multiply adherents, expand the organization, and lead operational and tactical missions and activities.

GOVERNMENT-IN-EXILE

2-91. A government-in-exile does not exist in every situation. A *government-in-exile* is a **government that has been displaced from its country, but remains recognized as the legitimate sovereign authority.** Whether a government-in-exile does or does not exist, the insurgency will usually still report to some form of a shadow government in-country. Governments-in-exile were very common in World War Two as Axis powers overran and displaced the preexisting legitimate government. These (French, Dutch, Polish, and so on) classic examples almost always represented legitimate government that gave sanction to resistance against occupiers within the home country. Today, governments-in-exile are most often associated with the remnants of preexisting governing authorities displaced by successful insurgencies, revolutions, coups, or occupation (the King of Afghanistan, and Burma, Tibet, or Baluchistan, for example). These are likely to be ineffectual and short-lived but may form the legitimate focus around which a resistance or insurgency might be organized.

SHADOW GOVERNMENT

2-92. A *shadow government* is defined as **governmental elements and activities performed by the irregular organization that will eventually take the place of the existing government. Members of the shadow government can be in any element of the irregular organization (underground, auxiliary, or guerrilla force).**

2-93. The shadow government is an organization the underground forms in occupied territory. Ideally, the shadow government can perform normal governmental functions in a clandestine manner and synchronize those functions with the resistance movement. The shadow government is critical because it exercises a degree of control, supervision, and accountability over the population at all levels (district, village, city, province, and so on), and further discredits and delegitimizes the existing government. The Chinese and Vietnamese communists are classic examples of effective shadow government while Hezbollah and the Taliban are recent examples.

POPULATION ("MASS BASE")

2-94. Although not traditionally considered a component of a U.S.-sponsored insurgent or resistance movement, the larger indigenous population from which the irregular forces are drawn must be an organizational consideration. UW is an elemental irregular warfare activity; as such, influence over a relevant population is critical. Communist insurgents of the 20th century referred to the general population as the "mass base," a hyperbolic image which magnified assumptions of popular support. However, the concept of the mass base had other more practical importance.

2-95. The communist method—especially since Mao—recognized the population as the raw material of revolution. Infiltration of popular institutions and organizations that could mobilize popular support was the key to influencing, recruiting, and manipulating ever-expanding popular groups. Insurgent leaders therefore tasked elements of the underground with infiltrating civil institutions and manipulating popular grievances and overt indigenous political activities to support insurgent objectives. Many of these activities, such as strikes, labor unrest, food riots, and so on may be effective methods of combating and weakening the adversary government, without being directly associated with the insurgent effort (Chapter 3 discusses techniques of these activities).

2-96. The participants in such mass-popular activities may be unaware of their manipulation by and for the insurgent movement. Therefore, planners and executors of UW campaigns cannot consider these participants a direct part of the insurgent organization. However, planners should take the insurgency's use of these participants in a U.S.-sponsored UW effort into account as an adjunct organizational factor.

AREA COMMAND

2-97. The **area command** is, in *unconventional warfare, the irregular organizational structure established within an unconventional warfare operational area to command and control irregular forces advised by Army Special Forces.* The area command denotes the resistance leadership that directs, controls,

Chapter 2

integrates, and supports all resistance activities within the AO. The area commander's location should be a place where he can safely control the insurgency and its activities. Flexibility, intelligence, mobility, and operations security are essential for survival and success. If the insurgency subdivides the area command into sector commands, its component units are the subordinate sector commands. The area commander is responsible for a broad region and typically there will only be one area commander. However, in very large-scale and/or widely dispersed insurgencies of resistance movements, it is possible that there is more than one region and more than one area commander. The area commander may or may not be the most senior strategic leader of the resistance or insurgency. Areas and area commands are typically subdivided into smaller sectors and organized under subordinate sector commands.

AREA COMPLEX

2-98. An area complex is a clandestine, dispersed network of facilities to support resistance activities in a given area. The area complex is contested territory or an area that contains clandestine supporting infrastructure. It is not liberated territory. It represents the insurgent's AO. Insurgent forces can maintain their clandestine infrastructure in the area complex. The clandestine infrastructure provides insurgent forces with a measure of freedom of movement and support. These areas overlay areas under the control of the government or occupying military. These areas can eventually transform into liberated areas if the enemy's ability to challenge the insurgent forces degrades to a level of parity with the guerrilla forces. To support resistance activities, an area complex must include a security system, guerrilla bases, communications, logistics, medical facilities, and a series of networks capable of moving personnel and supplies. The area complex may consist of friendly villages, towns, or portions of urban areas under guerrilla military or political control (Figure 2-7).

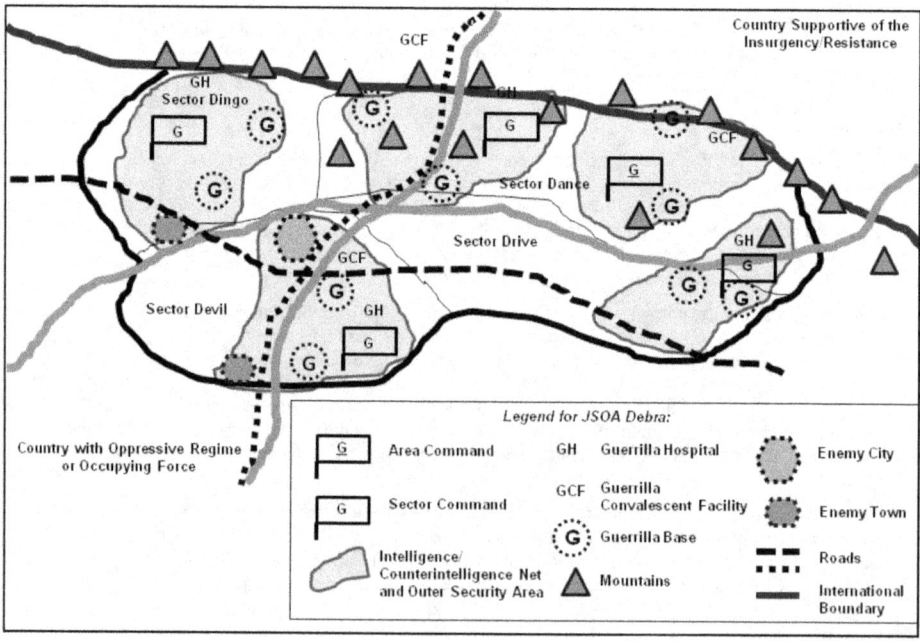

Figure 2-7. Joint special operations area "Debra" area complex sketch

Fundamentals

GUERRILLA BASES

2-99. A *guerrilla base* is a temporary site where guerrilla installations, headquarters, and some guerrilla units are located. A guerrilla base is considered to be transitory and must be capable of rapid displacement by personnel within the base.

2-100. A guerrilla base is an encampment that affords the guerrilla force the ability to rest, recuperate, plan, and train. Guerrilla bases may be temporary or semipermanent camps. Their level of complexity is proportional to the surrounding security situation. The security situation is a combination of the ability of the guerrillas to receive warning of an enemy advance, the ability of the enemy to project force to the base area, and the ability of the guerrillas to conceal their signature from all means of enemy detection. Guerrilla leaders should locate bases in terrain that counterguerrilla forces cannot easily access, such as restrictive rural or urban terrain. In environments that lack suitable restrictive physical terrain, certain urban environments can serve as guerrilla base locations. Densely populated urban enclaves that are sympathetic to the guerrilla cause can present an obstacle to counterguerrilla forces. Unlike actual restrictive physical terrain that always restricts the enemy's ability to attack, urban restrictive terrain only serves its purpose while counterguerrilla forces remain unwilling to assume the physical and political risk of entering the enclave. The security of the guerrilla camp comes from two bands of protection—inner and outer zone security—which planners refer to as bizonal security.

Inner Security Zone

2-101. The inner security zone encompasses normal camp security measures found in any military encampment. These measures consist of static guard and sentry positions, flank observation posts, and roving patrols. These measures provide a layer of physical protection and early warning to the potential surprise of an advancing enemy patrol or attack. The security elements have a coordinated means of signaling and communicating with the main base, as well as clear instructions that dictate how they will respond to different threats. Bases have emergency procedures to respond to enemy attacks. These procedures include plans for rapid evacuation and withdrawal to preplanned rally points and alternate base camps, as well as possible dispersal of the main body while a dedicated guerrilla element remains behind to temporarily delay any enemy penetration.

2-102. Fighting positions, obstacles, command-detonated or personnel-initiated mines, preestablished ambush positions, mortars, explosives, or traps along probable enemy vehicular and personnel avenues of approach may support the defense plans. It is essential that the guerrillas avoid becoming decisively engaged while carrying out their delaying and defensive mission.

Outer Security Zone

2-103. Insurgent leaders develop and organize the auxiliary's clandestine supporting infrastructure in the outer security zone. Clandestine lines of communication connect the guerrilla base and other facilities within the area complex. The outer zone consists of multiple networks of auxiliary agents that provide passive and active surveillance of key enemy positions that would indicate a pending enemy operation, such as airfields, motor pools, army compounds, police stations, or choke points along major roadways. These auxiliary agents might also be in positions to observe indicators of pending offensive operations, such as the daily itineraries of key leaders, the absence of clientele at popular enemy recreation spots, or the arrival of new special units. Auxiliary personnel may also be in positions to monitor unsecured telephone and radio traffic, collect and sort the enemy's trash, and take advantage of individual security force and soldier operational security awareness. Auxiliary members may also extort or intimidate enemy security personnel to procure operational information. Redundant collection methods and secure means of communications between auxiliary members and guerrilla bases provide the guerrillas with significant information that allows them to maintain the initiative over enemy security forces.

Chapter 2

INSURGENT SUPPORT NETWORKS

2-104. Just like other large organizations, insurgencies need support networks. These networks are usually run by either the auxiliary or underground, or a combination of the two. Insurgent support networks include the following:

- Logistics support networks.
- Transportation networks.
- Medical support networks.
- Communications networks.
- Information and propaganda networks.
- Recruitment networks.
- Intelligence and counterintelligence networks.
- Finance networks.

Logistics Support Network

2-105. Each resistance organization must develop a logistics system to meet the specific requirements of their situation. In general, however, a resistance organization meets its logistical requirements through a combination of internal and external means.

2-106. The area complex must provide the bulk of an insurgent organization's logistical requirements. The area commander must balance his support requirements against his need for civilian cooperation. Imposing excessive demands on the population may adversely affect popular support. Logistical constraints may initially dictate the size of the resistance organization.

2-107. As the resistance organization expands, its logistical requirements may exceed the capability of the area complex to provide adequate support. When this situation occurs, an external sponsor provides supplemental logistical support or the resistance organization reduces the scale of its activities. External support elements normally limit support to the necessities of life and the essential equipment and supplies the resistance needs to conduct combat operations. Internal sources of resistance supply include the following:

- Battlefield recovery.
- Purchase.
- Levy.
- Barter.
- Production.
- Confiscation.

2-108. Successful offensive operations permit resistance forces to satisfy some of their logistical requirements through battlefield recovery. Capturing supplies from hostile forces also avoids alienating civilians. The resistance organization normally limits its purchases to critical items unavailable by other means. Excessive introduction of external currency may disrupt the local economy, which may not be in the interest of the resistance organization or the United States.

2-109. The resistance organization may organize a levy system to ensure an equitable system for obtaining supplies from the local population. Levy systems manifest in a variety of ways from tithes and collections through sympathetic religious organizations to cross-border fees and formal taxes on goods and income. Under a levy system, the resistance organization provides receipts and maintains records of levy transactions to facilitate reimbursement at the end of hostilities. Obstacles to a levy system include—

- Chronic shortages among the local population.
- Hostile populace and resources control measures, including confiscation or destruction of local resources.

- Competition from the hostile power or rival resistance organizations.
- Chemical, biological, radiological, nuclear, and high-yield explosives or other contamination of local resources.

2-110. Barter may adversely affect the levy system. However, it is sometimes the only method of obtaining critical services or items, such as medical supplies.

2-111. Resistance forces often have to improvise their own field expedients. They may even have to plant and raise their own food, dig wells, and tend their own livestock. The area commander may consider establishing clandestine factories to produce unobtainable items.

2-112. Confiscation alienates the local population. The resistance organization should use confiscation only in emergencies or as punishment for individuals who refuse to cooperate or who actively collaborate with the hostile power. In all cases, resistance leaders must strictly control confiscation to ensure that it does not deteriorate into looting.

2-113. Guerrillas need the ability to acquire, store, and distribute large quantities of supplies without standard lines of supply and communications. They accomplish this by maintaining a decentralized network of widely distributed caches instead of large centralized stockpiles. This minimizes the loss of materiel if a guerrilla base moves quickly or faces destruction. This network allows the guerrillas to conduct operations across a wide area without a long logistics tail. The area commander caches extra supplies and equipment throughout the operational area. Caching is not a haphazard affair. Caches must support anticipated operational requirements or specified emergencies.

2-114. The logistics supply network also includes facilities for materiel fabrication, such as false documentation, improvised explosives and munitions, and medical aid. If the resistance is receiving external support, this network will extend to clandestine airstrips, drop zones, seaports, and border-crossing sites.

2-115. The resistance organization obtains repair materiel from the local economy and through battlefield recovery to perform all maintenance and repairs within its capability. It may establish repair facilities within the area complex. The sponsor includes necessary maintenance and repair items with all equipment it provides the resistance. Introducing sophisticated equipment into the area complex may complicate the maintenance system.

Transportation Networks

2-116. The resistance requires the capability to move personnel and logistics safely through enemy-controlled areas. Transportation networks include a compartmentalized series of safe houses or similar hiding locations. These locations allow the transport of personnel and materiel over long distances under the control of regional personnel who are familiar with the local enemy security measures. Security requires a complex series of recognition signals and communications that allow the individual segments to transfer the personnel and materiel safely with minimum exposure of either compartment to the other. These networks can also facilitate the evacuation of wounded personnel or personnel evading the enemy, such as downed airmen.

2-117. The area commander normally obtains transportation support from the auxiliary on a mission basis. The guerrilla force may have its own organic transportation system to meet its immediate needs. In remote or undeveloped areas, the primary means may be human porters or pack animals.

Medical Support Networks

2-118. The resistance initially confines clandestine medical treatment facilities to emergency and expedient care, with little preventive medicine. Once the area complex sufficiently develops, the clandestine facilities can expand and become a semipermanent medical organization, which serves the following purposes:
- To sustain and preserve combat power.
- To support the population.

Chapter 2

2-119. If the area command has not established a degree of clandestine medical support, the result will be evident in the guerrilla force's morale. Historically, a lack of proper medical attention has led to serious illness and disability that reduced overall unit effectiveness.

2-120. Medical elements supporting the resistance forces must be mobile, responsive, and effective in preventing disease and restoring the sick and wounded to duty. It is unlikely the movement will have a safe rear area where it can take casualties for treatment. Medical personnel help during combat operations by operating casualty collection points, which allows the healthy guerrillas to keep fighting. Medical personnel evacuate casualties from these points to a guerrilla base or civilian care facility.

2-121. The resistance organization requires basic medicines and other medical supplies to treat its members. Preventive medicine is especially important to a resistance organization because it normally does not have adequate facilities to treat diseases.

2-122. Resistance personnel use existing logistics and transportation nets to gain supplies and move casualties. The movement of wounded personnel across enemy-controlled areas by auxiliary members is a clandestine operation, not a support function.

2-123. There are three levels of care in the medical support network beyond simple self or buddy care. Each represents a progressively higher quality of treatment and usually represents a greater distance traveled within the system and away from the area of fighting. The three levels are—
- Aid station.
- Guerrilla hospital.
- Convalescent facility.

Aid Station

2-124. Aid stations are locations where trained medical personnel provide emergency treatment. Evacuation of wounded personnel from the battle area begins at these stations. Because the condition of the wounded may prevent movement to the unit base, personnel hide them in secure locations and notify the auxiliary. The auxiliary cares for and hides the wounded or evacuates them to a treatment facility.

2-125. The evacuation of the dead is important for security reasons. If the enemy identifies the dead, the families of the guerrillas may be in danger. Personnel evacuate and cache the bodies of those killed in action until proper burial or disposal of the bodies IAW the customs of the local population. Removal and burial of the dead denies the enemy valuable intelligence concerning indigenous casualties.

Guerrilla Hospital

2-126. A guerrilla hospital is a medical treatment facility (or complex of smaller facilities) that provides inpatient medical support to the guerrilla force. The resistance movement establishes a guerrilla hospital during the organization and buildup phase of its development. The hospital must be ready for operation at the start of combat operations and must be able to continue providing medical support until the leadership directs otherwise.

2-127. A guerrilla hospital rarely outwardly resembles a conventional hospital. The requirement for strict security, flexibility, and rapid mobility prevent visible comparison with conventional military or civilian medical facilities. As the guerrilla force consolidates its hold on the area complex, all medical support functions tend to consolidate. Safe areas allow the resistance to establish a centralized system of medical care. Sophisticated hospitals provide more elaborate care because they provide a wider selection of trained personnel and specialized equipment. These hospitals can also render more extensive and prolonged treatment.

Convalescent Facility

2-128. A convalescent facility is the area where guerrilla forces send patients to recuperate. A guerrilla convalescent facility may be a safe house in which one or two convalescents are recuperating with an appropriate alibi or it could be in any base in guerrilla-controlled areas.

Communications Networks

2-129. Guerrillas and underground leaders need to communicate with their subordinate elements in an area where enemy forces are always actively looking and listening for any indicators that would compromise the location of guerrilla forces or their supporting mechanism. Because of the likelihood of a high early warning threat, especially in the initial phases of the resistance movement, nontechnical communications should prevail.

Information and Propaganda Networks

2-130. Special networks are responsible for providing information to the population, against the will of the controlling regime. This information will bolster the will of the population to support the insurgent cause, undermine the legitimacy of the regime or occupying power, and undermine the morale of enemy security forces. Guerrilla forces may produce and distribute bootleg radio broadcasts, underground newspapers, Internet sites, social media postings/chat rooms, and rumor campaigns. Guerrilla propaganda networks also draw new recruits to the movement. The networks may also coordinate with sympathetic elements outside the country to raise international favor and support. The resistance or insurgent leadership must have a degree of communication with the propaganda network to produce a coordinated effort.

Recruitment Networks

2-131. The insurgency requires new recruits to join all aspects of the movement. The incorporation of these individuals requires special security measures to prevent the compromise of the components. The insurgency often sequesters recruits until it can check the recruit's validity and the recruit can complete training and possibly participate in an operation to prove his loyalty.

Intelligence and Counterintelligence Networks

2-132. Aside from normal intelligence collection requirements, the resistance must recruit new members. The resistance screens new members to ensure they are not infiltrators. Further details are beyond the scope of this publication.

Financial Networks

2-133. Financial networks have become a critical aspect of supporting, sustaining, and resourcing operations. UW planners must consider the protection of supporting financial networks while targeting the funding sources of the targeted regime/government. Financial support is typically drawn from a combination of domestic and international fundraising. Typical domestic sources include voluntary donations by supporters, tithes conflating religious duty with support for political activities, and legitimate money-producing front enterprises. Examples of involuntary domestic sources include collecting mandatory union dues and association fees that are then funneled into insurgent coffers, "revolutionary" taxes imposed directly on the populace, extortion from legitimate neutral businesses, and outright criminal activity, such as bank robbing or drug trafficking.

2-134. Insurgent financial networks can also enjoy a variety of international sources: direct donations from sponsoring or sympathetic states, sympathetic NGOs and purpose-built fundraising front groups, legitimate business enterprises, and diaspora communities. Whereas the primary emphasis of domestic financial networks is the clandestine accumulation, transfer, and expenditure of funds, the greater emphasis of international financial support is maintaining the covert nature of the sources through money laundering by banks and other legitimate businesses or series of front groups. In the information age, digital transference of enormous sums is often easily achieved. Meanwhile, ancient methods of illicit cross-border activity (such as smuggling) are as relevant today as ever.

This page intentionally left blank.

Chapter 3
Activities

Persuasion is often more effectual than force.

Aesop

It is imperative to acknowledge that what the United States refers to as "unconventional warfare" has been conducted by nations and groups throughout history worldwide, and most of the following activities—particularly political subversion of opponents—are fundamental to conducting UW. This U.S. Army publication is intended to instruct U.S. personnel on how to recognize and understand the types of techniques they may observe being used by others on the 21st century UW battlefield. The techniques explained below are therefore only a representative sample of what has been used historically by many actors. Presentation of subversive techniques in this publication is not a statement that U.S. forces necessarily conduct or support any one of them, nor should such U.S. conduct or support be inferred. The United States does not conduct or condone terrorism.

The following is a general unclassified description of the usual component activities of UW. More specific and detailed descriptions of many definitions below may be found in USSOCOM Directive 525-5, *(S/NF) Advanced Special Operations (U)*; USSOCOM Directive 525-16, *(S/NF) Preparation of the Environment (U)*; and USSOCOM Directive 525-89.

PREPARATION OF THE ENVIRONMENT

3-1. PE is an umbrella term for activities conducted in foreign countries to shape and prepare an area for potential operations. Operational environment is defined in JP 3-0, *Joint Operations*, as "a composite of the conditions, circumstances, and influences that affect the employment of capabilities and bear on the decisions of the commander." Since this composite will differ with each specific example of UW, PE activities will correspondingly differ with each specific campaign.

3-2. These activities normally precede—sometimes by years—the operations they are intended to support. PE activities are critical to the establishment, maintenance, and reestablishment of the necessary conditions whereby UW could be considered as a feasible USG strategic option. Detailed information on PE is located within FM 3-05.220, *(S/NF) Special Forces Advanced Special Operations (U)*, and Training Circular (TC) 31-16, *(S/NF) Special Forces Guide to Preparation of the Environment (U)*. When PE activities are conducted as part of a UW campaign, they are usually conducted as low-visibility operations, are almost always classified, and may be clandestine and/or covert in nature. These operations are defined as follows:

- *Low-visibility operations*: Sensitive operations wherein the political-military restrictions inherent in covert and clandestine operations are either not necessary or not feasible; actions are taken as required to limit exposure of those involved and/or their activities. Execution of these operations is undertaken with the knowledge that the action and/or sponsorship of the operation may preclude plausible denial by the initiating power. (JP 3-05.1, *Joint Special Operations Task Force Operations*)
- *Clandestine operation*: An operation sponsored or conducted by governmental departments or agencies in such a way as to assure secrecy or concealment. A clandestine operation differs from

Chapter 3

a covert operation in that emphasis is placed on concealment of the operation rather than on concealment of the identity of the sponsor. In special operations, an activity may be both covert and clandestine and may focus equally on operational considerations and intelligence-related activities. (JP 3-05.1)
- *Covert operation*: An operation that is so planned and executed as to conceal the identity of or permit plausible denial by the sponsor. A covert operation differs from a clandestine operation in that emphasis is placed on concealment of the identity of the sponsor rather than on concealment of the operation. (JP 3-05, *Special Operations*)

SUBVERSION

3-3. A hostile government or occupying power relies on some critical minimum degree of military infrastructure and economic, psychological, and political strength and morale. Insurgencies and resistance movements attempt to undermine the human component of such strength and morale by employing techniques of subversion. Of all component UW activities, the subversive efforts of U.S.-supported insurgencies and resistance movements is the single most strategically valuable activity because it provides discrete methods of influencing an opponent's behavior without resorting to more overt, large-scale, and unilateral U.S. actions. U.S. support to "subversion" conducted by indigenous actors is the very heart of UW. Subversion, subversive political action, and psychological action are defined as follows:
- *Subversion*: Actions designed to undermine the military, economic, psychological, or political strength or morale of a governing authority. (JP 3-24)
- *Subversive political action*: A planned series of activities designed to accomplish political objectives by influencing, dominating, or displacing individuals or groups who are so placed as to affect the decisions and actions of another government. (JP 1-02, *Department of Defense Dictionary of Military and Associated Terms*)
- *Psychological action*: Lethal and nonlethal actions planned, coordinated, and conducted to produce a psychological effect in a foreign audience. Also called PSYACT. (FM 3-53, *Military Information Support Operations*)

Note: The detailed knowledge required to understand the vulnerabilities of a nation's political, economic, military, and social organizations and the sophisticated analysis of target audience susceptibilities is fundamental to the successful conduct of subversion and UW. Early inclusion of, and continuous coordination with, PSYOP and CA Soldiers in UW campaign planning is critical.

3-4. Typically, insurgencies are won by a combination of military and political means. Much of the political leverage involved in settlements is derived from the psychological and practical effects of mobilizing the populace against the government. Insurgent subversion strategy is to separate the existing government from its basis of power by capturing the institutional supports upon which it rests, by eroding mass support for the government, and by overtaxing internal security forces with problems of unrest. Insurgencies may not be able to automatically seize power through fomenting disorder, but the undermining of regime power and the creation of a security and governance vacuum is a prerequisite to the insurgency establishing its own control. When obedience to law breaks down among a populace, a tense, highly emotional state develops, which gives the underground a chance to channel dissatisfactions. Insurgent subversion methods and tactics involve the psychological objectives of creating social disorganization, and the political objectives of creating alternative organizations to contend for power. Psychological objectives may include the following:
- Create social disorganization and conditions of uncertainty. The resultant unrest and confusion are used to conceal underground operations. A characteristic of this kind of social confusion is a condition of general apathy among a large segment of the populace and an unwillingness to help either side. This indifference benefits the resistance underground; apathetic people do not cooperate by supporting government programs and they seldom volunteer the intelligence information necessary for detecting underground elements and operations.

- Create doubt and suspicion of government and government officials. This focuses attention and grievances on the ineffectiveness of government.
- Crystallize attitudes and organize dissident elements to resist government action and policies.

3-5. Subversive influence activities not only affect the operational environment in which populations reside, but the informational environment as well. The "real" world or the facts are relatively unimportant in subversion operations; the shaping of attitudes, values, and beliefs are vital, however. Underground organizations align their appeals with a society's recognized, accepted values—such as "independence" and "land for the landless." Members of the population who already accept widely held values are more easily persuaded to accept an insurgency which espouses them. Insurgents also reward those who are "loyal" and punish any who oppose them. Riots and passive resistance provide strong social pressure to influence the undecided or uncommitted.

3-6. Subversive influence activities are conducted in a number and variety of forms: PSYACTs; mass media; interpersonal communication; theatrical performances; programs for local civic improvement; and audio, visual, and audiovisual messages. Although the themes, targets, objectives, and other informational guidance used throughout the operation is determined at the highest echelon of the organization, successful execution of influence activities depends in large part upon the training, experience, and ingenuity of the operators at the local level.

3-7. In attempting to shape popular attitudes, values, beliefs, and behavior, and to develop support for subversion activities, influential messages and actions are directed at specific target audiences or groups. Occupational, religious, ethnic, and other social groups are often identified, and messages and actions are tailored to be effective for a particular group. The purpose of underground information activities may be to obtain support from neutral and uncommitted key groups and individuals; to raise morale and reinforce existing attitudes and beliefs among underground members and their supporters; to undermine the legitimacy and credibility of the existing government; and to lower the morale of government forces and personnel.

3-8. Underground movements tailor emotional and other types of appeals to various segments of society. For example, groups that are reluctant to take up arms against the government may be susceptible to emotional appeals and directed into passive measures. Religious or pacifist groups, women, children, and the elderly may also possibly be mobilized for passive resistance. In organizing demonstrations and riots, attention is given to selecting groups most likely to respond to the agitator's call to action—student groups, dissatisfied labor union members, and groups with known grievances. Insurgencies and resistance movements can use subversive techniques to manipulate crowds and civil disturbances for the purpose of advancing the overall insurgent strategy.

INFORMATION ACTIVITIES

3-9. Resistance organizations deliver influential messages to audiences while agitation typically consists of interpersonal and task-specific communication directed toward relatively small, selected audiences. Messages and action (agitation) are interdependent and complementary. Typically, mass communications build support for the cause of the movement by articulating emotions associated with conditions (grievances) through messages, and agitation "operationalizes" those messages by associating the movement with actions being taken to solve or address those grievances on behalf of the people affected.

MESSAGES

3-10. Persuasive messages attempt to create feelings of doubt and uncertainty about future events and to promote perceptions of crisis. Care is taken to differentiate between the government and the people, and attribute blame or fault to the government. Message audiences are generally divided between broad, general ones based on demographics, region, and so on, and those crafted for more local reception.

3-11. Messages typically used by resistance organizations consist of three types: informative, directive, and persuasive. When the message captures people's attention, they will be receptive to the persuasive content of the insurgent's message.

3-12. In order to ensure continuity and consistency, selected messages are provided at regular intervals to all resistance information organs. In addition, insurgencies and resistance movements may have their own doctrine and procedures that must be trained and reinforced. The larger, more mass-based, and more multinational an insurgency is, the more this will be true. Moreover, such "instructional" content may be used for execute orders to dispersed resistance cells. Previous insurgencies have used the "double language" routine, where disguised instructions are embedded in routine media broadcasts.

3-13. Persuasive messages are designed to bolster the morale of the insurgents, to undermine confidence in the government and its policies among audiences, and to win active supporters or at least sympathizers to the movement. The messages tend to be phrased in highly emotional terms.

General Themes

3-14. Several devices are used to justify the movement through a "consensual validation"—creating the appearance of majority approval. Resistance themes tend to stress the legitimacy as well as the reality of insurgent power vis-à-vis the opposing regime. Some examples include the following:
- Insurgents speak for the people.
- The opposing regime is ruthless, engaging in unwarranted and unprincipled aggression to exploit the populace.
- The opposing regime is arrogant and contemptuous of the indigenous population.
- The opposing regime is deceitful in all actions and statements; nothing done by the regime is as it seems.
- The power of the opposing regime is overstated; it is not to be feared.
- The victory of the insurgency is inevitable.
- External support enjoyed by the insurgency is framed as widespread and legitimate international support, whereas regime external support proves collusion with foreign interests, betrayal of local interest, and lack of regime popular support.
- In some cases, the prevailing theme is the effectiveness of individual initiatives taken against the government, whereas other themes stress the helplessness of individuals who ought to bind themselves to the collective strength of the insurgent movement.

Local Appeals

3-15. Appeals can facilitate insurgent organization building, or at least encourage acquiescence with insurgent activities. Some local target audiences may be urbane and politically motivated enough to be persuaded by theoretical arguments. These, however, are the exception. Most target audiences will be initially reluctant to participate in anti-regime activities and will be most persuaded by local issues of grievance and security. Local appeals are characterized by—
- Avoiding theoretical arguments versus demonstrated provision of improving local goods and services.
- Couching allegiance in terms of who will eventually win and who will lose rather than consideration of who in the long run is right or wrong.
- Exploiting family and community ties, and local preferences and prejudices.
- Motivating and serving the populace (insurgent emphasis) versus eliminating insurgent fighters and repressing the populace (regime emphasis).
- Demonstrating regime impotence and absence versus insurgent effectiveness and sustained presence (action propaganda).

Mass Media

3-16. There are several techniques of message delivery. The following paragraphs discuss these techniques.

Television

3-17. The television has been an important means of mass message delivery—especially in the developed world—for well over half a century. As globalization occurs, the reach of television expands concurrently. Wikipedia estimates that the number of television sets worldwide in 2012 is approximately 1.4 billion, or 1 television set for every 5 people on earth. Other than cell phones, television is the largest source of unified messaging technology available. Moreover, television content transmission is expanding due to increasingly widespread use of satellite signals replacing earth-bound line-of-sight transmissions.

Social Media

3-18. The widespread use of cell phone and Internet technology allows agitation on an unprecedented scale. These social media are much more responsive than a leaflet, radio, or newspaper and other mass media, and when signals are unblocked by the government, they allow real-time organization across wide distances and among disparate groups. Wikipedia estimates that the number of cell phones on earth is approximately 5.6 billion. This means the number of cell phones worldwide is approaching the number of human beings.

Newspapers

3-19. Newspapers are important messaging tools. Newspapers that are centrally controlled help spread insurgent information and ideas uniformly and so are instrumental in tying its readers into a close mental community; they become a mental rallying point—they "spread the party line." However, the importance of such clandestine newspapers extends beyond the obvious function of influence. Production and distribution of subversive newspapers is an organizational challenge that contributes to resistance organization building.

> *Arranging for and organizing the speedy and proper delivery of literature, leaflets, proclamations, etc., training a network of agents for this purpose, means performing the greater part of the work of preparing for future demonstrations or an uprising.*
>
> V.I. Lenin

3-20. Undergrounds not only produce their own newspapers, but also use existing legitimate newspapers for their own ends. Press criticism of certain aspects of local authority can provide fertile ground for implantation of rumors about public officials targeted for subversion. Insurgent undergrounds will seek to infiltrate established media entities to influence their coverage choices and alter their editorial positions.

Radio

3-21. Radio broadcasts have the obvious advantage of simultaneously reaching a large number of persons over a considerable range of territory. The same coverage by newspaper takes longer and is much more dangerous for the publishers, distributors, and recipients. Radio broadcasts can be made abroad, providing challenges to enemy efforts to shut them down. Likewise, foreign-based insurgent radio stations can claim to be broadcasting from within the "liberated territory," a claim which is both embarrassing to the government and difficult to disprove.

Interpersonal Communications

3-22. Possibly the oldest method of passing information is also one of the most effective: "word-of-mouth" interpersonal communications. Resistance movements or insurgencies operating in rural, agrarian, low-literacy societies can effectively influence and shape local attitudes through word-of-mouth transmission.

3-23. In some cases, travelers such as merchants, itinerant storytellers, religious figures, or dedicated and professional resistance cadres can make circuits of remote areas. Many remote villagers are likely to be both illiterate and lacking outside sources of information. Such persons are vulnerable to influence by skilled subversives. The remoteness of outside information is easily discredited or reinterpreted toward insurgent goals by agents who physically visit the remote area as part of a larger coordinated plan of continuous persuasion.

3-24. Some mass-based insurgencies will develop village infrastructures with dedicated local political agents to control access to information. Such persons may control the only portable radio, cell phone, or newspaper in the village. The agitator, to ensure a maximum degree of credibility, tries to bar other external sources of information from his target group or area. This can be done through the confiscation of radios, threats to rural newspaper distributors, and impairment of government access to the area. In effect, the political agent seeks to develop a captive audience in order to facilitate his job of influencing attitudes and behavior and generating popular support for the underground movement.

AGITATION

3-25. Agitation is essential for creating mass support. It takes more than a presentation of information to a group of people for them to accept a cause and be persuaded to support it. Exposure to information does not imply absorption of it. There are psychological as well as physical barriers that inhibit the flow of information and ideas. There may be general apathy in which a large portion of the population is unfamiliar and unconcerned with particular events. Another barrier is the phenomenon of "selective exposure": a tendency to hear only information conforming to individual tastes, biases, and existing attitudes. There is also "selective interpretation": information understood only in terms of current attitudes. Frequently, individuals who do alter their attitudes as a result of new information do so only within the context of their prior attitudes.

3-26. The task of the agitator lies in overcoming these barriers by developing and delivering messages that are credible and meaningful. Agitators must reach the indifferent; must blend messages with the existing attitudes of target groups; and must make the resultant attitude change that can be exploited to elicit mass action. The agitator must remove any complacency that exists among a group of people, intensify their unrest, and channel it to suit the purposes of the underground.

PASSIVE RESISTANCE

3-27. People frequently overestimate the effectiveness, centrality, and importance of guerrilla warfare to UW, and tend to underestimate the critical role of subversive populace organization and the time required to do such organizing. Resistance movements or insurgencies whose strategy relies on widespread involvement of the population must be able to mobilize the population to engage in widespread passive resistance. In UW, it is primarily operatives of the underground who organize and direct passive resistance efforts to persuade the ordinary citizens to carry them out.

Objectives

3-28. Passive resistance implies a large, unarmed group whose activities capitalize upon social norms, customs, and taboos in order to provoke action by security forces that will serve to alienate large segments of public opinion from the government or its agents. If the government does not respond to the passive resisters' actions, the resisters will immobilize the processes of public order and safety and seriously challenge the writ of government.

3-29. Passive resistance rests on the basic thesis that governments and social organizations, even when they possess instruments of physical force, depend upon the voluntary assistance and cooperation of great numbers of individuals. Therefore, the passive resistance method of opposing an established power structure is to persuade as many persons as possible to refuse to cooperate with it, including members of governmental organizations, such as security and the military.

3-30. The principal tactic used to induce noncooperation with the government can be described as "persuasion through suffering." One of the persistent myths of passive resistance is that persuasion through suffering aims only to persuade the governing opponent by forcing it to experience a guilty change of heart and a sense of remorse. This conception of the role of suffering makes the fundamental error of presuming that only two actors are involved in the process of passive resistance—the government and the resisters.

Actually, passive resistance operates within a framework involving three actors: the government, the resisters, and the critically important larger public "audience" (the population in general). More than anything else, the objective of passive resistance is to create situations that will engage public opinion and increase popular involvement in resistance activities, and to wear down and frustrate the overall effectiveness of the established power structure.

3-31. When the passive resister suffers at the hand of the government, the insurgent seeks to shape the interpretation of that suffering as a demonstration of resister integrity, commitment, and courage, while simultaneously demonstrating the injustice, cruelty, and tyranny of the government. The essential function of such suffering is comparable to the emotional mob interaction that takes place between a martyr and a crowd.

3-32. If the passive resister provokes a response from the security forces or government which can be made to seem unjust or unfair, the resister's charges of tyranny and persecution are confirmed. Should the government fail to act, it abdicates its control over the population, over the enforcement of law, and over the maintenance of order. This dilemma thrusts upon a government the initiative, and also the responsibility, for uninvited conflict with unarmed citizens. Either way, the position of the passive resister is legitimized, the effectiveness and morale of government security forces is degraded, and the insurgency's subversive political strength is increased. The ideal situation is where popular passive resistance is self-generating, self-multiplying, and ultimately victorious with relatively little violence as in India in the 1940s or in both Poland and the Philippines in the 1980s.

Techniques

3-33. Actions of passive resistance may range from small isolated challenges to specific laws, to complete disregard of governmental authority. In all cases, passive resistance is a two-edged sword: it seeks to lower the morale and effectiveness of the governing regime or occupying power, while at the same time raising the morale of the populace and creating a feeling of defiance and unity which could be channeled later into more significant resistance activity. The techniques of passive resistance can be classified into three general types: attention-getting devices, noncooperation, and civil disobedience.

Attention-Getting Devices

3-34. Passive resistance in the early stages usually takes the form of actions calculated to gain attention, provide persuasive messages for the cause, or be a nuisance to government forces. Attention-getting devices include demonstrations, mass meetings, picketing, and the creation of symbols. Demonstrations and picketing help advertise the resistance campaign and educate the larger public to the issues at stake. Such activities provide information and agitation for both internal and external consumption.

3-35. Many insurgencies have used massed demonstrations and general strikes—notably during the Iranian and Bolshevik revolutions—to provoke regime repression and bring attention to the cause. World War II (WWII) resistance efforts provide several examples of both nuisance campaigns (spitting on occupiers, cold-shoulder ostracism, deliberate rudeness, deliberate joke-and-ridicule campaigns, "whispering" rumor campaigns, and false reporting and over-reporting of bogus threats) and creation of symbols (the aforementioned martyrs, the cross of Lorraine, graffiti slogans, or the continued and very public presence of the Danish king as a symbol of continued Danish identity under Nazi occupation). Possibly the most dramatic example of symbol creation were the very effective self-immolations in Vietnam in the 1960s and before the "Arab Spring" in 2010–2011.

Noncooperation

3-36. Techniques of noncooperation call for a passive resister to perform normal activities in a slightly contrived way, but not so that police or the government can accuse him of breaking ordinary laws. Such activities as "slowdowns," boycotts of all kinds, and various forms of disassociation from government are all examples of noncooperation. Other examples include falsification of blueprints so that structures are built improperly; deliberate errors in adjustment of machine tools and precision instruments that creates flaws in end items; shipping mail or parcels to wrong addresses, shipping incorrect orders, or "forgetting"

Chapter 3

to include items in the shipment; feigned sickness which leaves gaps in the workforce; feigned fear which provides a rationale for both poor work performance and attendance, and taxes the health system; and many other examples. Such techniques of noncooperation can also be considered sabotage.

3-37. Noncooperation is a principal tool of passive resistance and has been shown to be most effective in disrupting the normal processes of society and severely hampering and challenging the writ of a government—all in a way that is difficult for the government and its security forces to challenge. Many individuals altering their normal behavior only slightly can add up to a society behaving most abnormally.

Civil Disobedience

3-38. Mass participation in deliberately unlawful acts—generally misdemeanors—constitutes civil disobedience. This is perhaps the most extreme weapon of passive resistance; the boundary between misdemeanors and serious crimes can be considered the dividing line between nonviolent and violent resistance.

3-39. Examples of civil disobedience include the breaking of specific laws, such as tax laws (nonpayment of taxes); traffic laws (disrupting traffic); and laws prohibiting meetings, publications, free speech, and so on. Civil disobedience can also take the form of certain kinds of strikes and walkouts, resignations en masse, and minor destruction of public or private property.

3-40. Civil disobedience is a powerful technique, but to be effective it must be exercised by large numbers. There is a calculated risk involved: the breach of law automatically justifies and involves punishment by the government and security forces. However, the more massive the scale on which civil disobedience is organized, the less profitable it is for the government to carry out sanctions.

3-41. Organizers of passive resistance are selective about the laws that are to be broken. The laws should be related in some manner to the issues being protested or the demands being made. In summary, the underlying consideration in most passive resistance techniques is whether they serve to legitimize the position of the passive resister while alienating or challenging the government.

Organization

3-42. Obviously, the success of passive resistance rests largely on its ability to secure widespread compliance within the society. A government cannot be robbed of the popular support upon which it depends if only a few individuals act. A boycott, for example, requires participation by great numbers.

3-43. Organization is of critical importance to passive resistance. Although a few individuals can launch a passive resistance movement, in order to succeed they must be joined by thousands whose participation is strategically channeled. Modern technology has made the organization of large numbers of sufficiently motivated resisters in a short period of time much easier than in previous eras. Given proper cadre planning and coordination, the summoning and mustering of resistance "flash mobs" at strategic locations can have potentially decisive psychological and political impact. Support to the indigenous underground in developing, growing, and providing both tactical and strategic employment of such popular resistance organizations is an early, continuous, and critical focus of U.S. Army SF conducting UW. Once again, the detailed social, demographic, and psychological expertise of U.S. Army CA and MISO units is invaluable in planning for such passive resistance, and for identifying what forces and factors induce people to mobilize for passive resistance.

Normative Factors

3-44. One method by which leaders of passive resistance movements secure widespread compliance is by clothing their movement and techniques in the beliefs, values, and norms of society—those things people accept without question. For example, when a minority secular regime like the Shah's or a minority secular occupier like the Soviets attempted to force social change on majority conservative Islamic societies in Iran and Afghanistan, such initiatives provided a ready-made focus for organizing popular resistance. They offended the traditional and dominant norms.

3-45. An opposite example would be constriction or repression of individual liberties in a society accustomed to enjoying them. The American Revolution and all of the multinational resistance to Nazi occupation in WWII were in part reactions to lost liberties. Other examples include the nearly universal disgust at the abuse of women, children, the elderly, beloved church leaders, and so on.

Consensual Validation

3-46. The technique of "consensual validation"—in which the simultaneous occurrence of events creates a sense of their validity—is often used to make public opinion coalesce. For example, if demonstrations take place at the same time in diverse parts of a country, the cause which they uphold appears to be valid simply because a variety of persons are involved. A minority group can organize a multitude of front organizations, so that seemingly widely separated and diverse organizations simultaneously espouse the same cause and give the impression that a large body of opinion is represented. Passive resistance organizers effectively use the psychology of consensual validation to rally public opinion.

Religious Factors

3-47. Rare or extraordinary factors such as charisma play an important part in mobilizing public opinion. Obviously, exceptional resistance leaders like Gandhi and Martin Luther King provided probably indispensable charismatic power to mobilizing public sentiment. Polish-born Pope John Paul played a significant role in solidarity's success, as did Ayatollah Khomeini in Iran, and the self-immolating Buddhist monks in Vietnam. In all three cases, the difficult-to-quantify factor of religious authority contributed to mobilizing passive resistance. Graphic Training Aid (GTA) 41-01-005, *Religious Factors Analysis*, includes more information on conducting analysis to better understand religious factors in planning resistance.

Pressure for Conformity

3-48. The same techniques used by passive resisters against the government can be used to ensure widespread social compliance within the resistance movement. Ostracism is frequently used to apply pressure on individuals not participating in the passive resistance campaign. Instances of organized ostracism of collaborators were common from occupied territories in Europe and Asia during WWII through resistance to colonial powers throughout Africa in the late 20th century.

COMMUNICATION AND PERSUASION

3-49. As noted earlier, the first phase of passive resistance is characterized by a period of attention-getting information activities and PSYACTs: parades, demonstrations, posters, newspapers, and other forms of communication, either clandestine or open. Once the resistance movement is launched, there must be continuing means of "spreading the word." No movement can operate without some form of communication between the leaders and the led.

TRAINING

3-50. Once organizational steps are taken to secure widespread social compliance, an effort must be made to instruct and train passive resisters. The idea is to erect a mental barbed-wire fence between resisters and authority. This instruction often takes the form of codes of "dos and don'ts." Many undergrounds have found that it is easier to tell people what not to do than what to do.

3-51. Training is particularly critical when positive, not just negative, actions are desired. Noncooperation and civil disobedience are positive acts that necessarily involve training, organization, and solidarity on the part of the resisters, whether they operate in the open or clandestinely.

PARALLEL GOVERNMENT STRUCTURE

3-52. One method that is frequently used to both undermine public confidence in a government and secure population support for passive resistance efforts is the establishment of parallel structures of government.

If a population depends upon an underground-sponsored government, it will be compelled to comply with the underground's passive resistance program and withdraw its support from the regular government.

3-53. The techniques and societal values capitalized upon to undermine popular support of the government also serve the positive function of solidifying public opinion around a larger sense of community and national identification. Examples of constructing parallel governments include Hezbollah in Lebanon, the Naga in India, the Fuerzas Armada Revolucionarias de Colombia (FARC) in Colombia, and the classic communist template from China to Cuba to Vietnam.

THE SUBVERSIVE MANIPULATION OF CROWDS, RIOTS, AND DEMONSTRATIONS

3-54. The next step—more dramatic, sometimes violent, and notoriously difficult to control—is the wielding of popular support as a weapon. Crowds can be subversively manipulated to conduct peaceful mass political gatherings, aggressive demonstrations, or riots that effectively support the insurgency or resistance. Popular unrest in the form of strikes, riots, and demonstrations usually have limited goals, such as better working conditions, relief from food or other shortages, or demands for limited social change. Typically, these activities are relatively spontaneous, narrowly focused, and short-lived; a single peaceful mass political demonstration or riot does not make an insurgency.

3-55. The goal of the insurgent underground, however, is the overthrow of the government and the seizure of power. The insurgent does this by either exploiting such spontaneous grievances in the service of the resistance movement, or by deliberately creating such demonstrations. The subversive manipulation of crowds and civil disturbances involves a relatively small number of insurgent underground members who try to guide and direct legitimate protests. They attempt to direct the crowd toward emotional issues and arouse them against authority. The emotional perceptions and beliefs of the crowds that participate in civil disturbances often do not coincide with objective reality, and the individuals involved very often do not realize that their grievances are being manipulated by the insurgency or resistance movement in politically subversive ways.

PHASES OF SUBVERSIVE MANIPULATION

3-56. Subversively manipulated civil disturbances may be considered as having four phases:
- The pre-crowd phase.
- The crowd phase.
- The civil disturbance phase.
- The post-civil disturbance phase.

Pre-Crowd Phase

3-57. In the preparation or pre-crowd phase, the underground elements are primarily concerned with building an organization. Training the network of agents necessary for the rapid and appropriate delivery of informative and persuasive messages accounts for the greater part of the work required to prepare for a demonstration or uprising. This step is a necessary precursor to strike or demonstration organization.

3-58. Selected individuals are given special training in the manipulation of crowds. They are taught how to build barricades and conduct street fighting, how to mobilize blocks in the city and workers in plants, how to develop a local strike into a general strike and a general strike into a city uprising, and how to coordinate these into a national uprising. Sometimes, outside specialists can be brought in to direct the training activities. They are instructed in the potential for nonviolent demonstrations and specific techniques for sit-ins, no-shows, silent mass gatherings, marches, public mourning for dead compatriots, suspension of civic and social activities, economic boycotts, refusal to pay taxes or fees, traffic shut-downs, and many other forms of resistance that provoke government reaction.

3-59. Some sort of planning on the part of the underground must take place. It may vary from rudimentary to highly sophisticated. Underground agents are instructed to infiltrate target groups by joining formal organizations, clubs, or any association which gives them access to such audiences.

3-60. The next step is the selection of a population target. It is chosen primarily for its potential to achieve the desired psychological or political objective. Furthermore, any group that is not susceptible to manipulation, at least after some preparation, is not considered a viable target. Groups identified by their common interests (for example, ethnic minorities, laborers, farmers, educators, youth, the unemployed, and so on) offer great potential for covert manipulation because attention can be centered on issues affecting them rather than on philosophical or ideological arguments.

3-61. The desired change in the attitudes and behavior of the members of the target group is usually accomplished through the delivery of carefully crafted messages and executing PSYACTs, the desired effects of which are intended to increase anxiety and emotional stress. Word of mouth, radio, telephone, leaflets, and other messaging have been effectively used. Pistols, rifles, materials for making Molotov cocktails and explosives, and other weapons, such as clubs and lengths of pipe, and also handbills, signs, armbands, and banners, must be acquired and stored.

3-62. Arrangements for members of the underground group to flee the area must be completed. These consist primarily of establishing routes of escape containing safe houses or other hiding places. Safe zones are established with householders and shopkeepers where demonstrators may seek cover when fleeing from the police.

3-63. In places where demonstrations or strikes can be planned in advance, the underground mounts a campaign directed at preconditioning target groups. Chosen themes are constantly emphasized in messages and actions. By concentrating on local and specific grievances, a group is conditioned to phrases and slogans to which its members may later react under conditions of emotional stress.

Crowd Phase

3-64. The indispensable element in civil disturbances is the crowd—not just any crowd, but a crowd made up of individuals who have been conditioned either by subversive manipulation or by other events. There are several ways to assemble a crowd. Cell members infiltrate mass organizations so that strikes or mass meetings can be changed into armed demonstrations. There are built-in sanctions within labor unions or other disciplined organizations that can be used to punish members who do not comply with the decisions of the organization. Therefore, if the infiltrated union or organization calls for a strike or demonstration, its members can be brought into a particular place at a particular time. Student groups are highly volatile on many social issues and can be induced to participate in demonstrations for the sheer excitement. Some factors of crowd formation and subversive manipulation include—

- *Informal gatherings*. Demonstrations can be brought about at parades, street parties, dances, or during normal rush-hour periods.
- *Hired demonstrators*. Organizers bring in paid outside agents and ruffians to expand crowds and accelerate radicalization.
- *The precipitating event*. The precipitating event which results in the formation of a crowd depends for a great deal of its effectiveness upon communication, especially upon exploitation of the event. Organizers associate the event with existing issues or exploit the natural distortion which accompanies word-of-mouth communications. The precipitating issue or event can be a martyred individual, a report of police brutality, or a symbolic act such as the desecration of a flag.
- *Mob management techniques*. Mob management has several general characteristics, to include—
  - Leader elements are organized into an external command—well-removed from the activity and which can observe the demonstration (and outside reaction)—and an internal command located in the crowd, which is responsible for directing the demonstration.
  - Bodyguards who surround and shield the internal command from the police and facilitate their escape, if necessary.
  - Messengers who carry orders between the internal and external commands.

Chapter 3

- Shock guards who are armed with weapons and act only as reinforcements or are able to create diversionary violence allowing resistance leaders to escape if they became engaged by the police.
- Banner carriers who switch from banners expressing general grievances to those reflecting direct insurgent propaganda at the appropriate time.
- Cheering sections consisting of special demonstrators who rehearse the slogans and chants and the order in which they are to be raised.
- Media handlers who ensure that sympathetic media are well-placed to capture dramatic and inflammatory images useful to insurgent propaganda.

- *The agitator.* After the crowd has been formed, the agitator assumes a significant role. His function is to enflame smoldering resentments of his listeners through emotional appeals and then provide a justification for their resistant actions. The agitator in the crowd plays upon the audience's suspicion of things they do not understand and manipulate whatever grievances motivate that specific crowd (material deprivation, repressive laws, inequality, nationalist or religious redress, martyred comrades, or other perceived injustices). The agitator then further enflames the crowd by painting a mental picture of dangers, disasters, insecurity, uncertainty, and hopelessness. Critically, the agitator must then identify the "other"—the bogeyman or oppressor responsible for the injustices and the conditions and who must now be opposed by that particular crowd. Where the conditions for insurgency or resistance are latent and incipient, the agitator rarely needs to invent grievances, merely exploit them. Even when issues are contrived, the skilled agitator rarely justifies his facts, but relies on vague appeals, generalities, and emotional themes.
- *The disturbance leader.* After the crowd has been emotionally aroused, some event must set it in motion. Often it begins its riotous activity by following a leader who merely shouts "Follow me! Let's go!" The insurgent internal command assumes the leadership role if an emergent leader does not arise spontaneously. The event which sets the crowd in motion may, like the precipitating incident that brought them together, be either factual or fabricated. In the case of nonviolent protest or strike, the disturbance leader will act to calm the crowd in the face of opposition from the security apparatus or opposing groups. The disturbance leader will ensure that heavy-handed repression by the government is witnessed and recorded so that it can be exploited to build local, national, and international support.
- *Commands, signals, and communications.* Small groups have face-to-face communications and interaction. In large groups, however, communications come through second- and third-hand sources. Insurgent media, preplanned distributions, cell phones or tactical radio, social media devices, and so on, all aid the magnification of insurgent command and control. A relatively new political phenomenon is the "virtual crows," which presents the agitator with a significant opportunity to proselytize and preprogram wide audiences who are then subject to rapid mobilization and exploitation. Very large actual crowds—or many crowds dispersed over a large area—can also be susceptible to rumors. This was truer before the widespread use of social media. However, social and other electronic media can be blocked rendering the crowd again vulnerable to word-of-mouth influence. Moreover, skilled operatives can use social and other electronic media itself to spread rumor campaigns. Rumor can be useful to mobilize collective action because they float on generalized beliefs rather than on demonstrable facts, and tend to lend perceived substance to generalized beliefs. Where such generalized beliefs can be identified early, or even pre-planted into mob consciousness, agile insurgent use of timely rumors can spark desired crowd reactions.

Civil Disturbance Phase

3-65. Some factors of shaping and controlling crowds once they are rioting include—

- *Maintaining emotional excitement.* Once the destructive action of the crowd is under way, the agitator tries to maintain the level of emotional excitement. This can be accomplished in various ways. Cheerleaders can chant rhythmic and inspiring phrases or songs. Slogans can be displayed and banners unfurled. "Booster" incidents—the most universal being the looting of stores and shops—can be created or capitalized upon. Bank holdups and kidnappings are also carried out during the chaos. Other acts, such as the verbal abuse and stoning of police, not only permit the individual to release aggression and hostility against the symbols of authority but also increase the emotional involvement. The use of Molotov cocktails or similar dramatic destructive events also ratchet-up crowd excitement.

- *Expanding the violence.* Agitators attempt to capitalize upon the contagious effect of civil disturbances by spreading the violence and creating new incidents in nearby areas. Attacks upon symbols of authority, such as police stations and the offices of local officials, increase the intensity of the disorder. If possible, radio stations, newspapers, water, and power services are seized. Newspapers and radios spread the rumors, and control of water and power plants spreads social disorganization and fear.

- *Counter-police activity.* Police and army counter-riot tactics are studied by the insurgent planners so that steps can be taken to circumvent them. Routes usually taken by internal security forces are blocked with barricades, overturned vehicles, and debris. Attacks upon police stations and their communications systems serve to disrupt police countermeasures. Cadres are usually guarded by strong-arm squads and avoid confrontation with the police so they will not be jailed. Appeals are made to army or police units not to attack their own countrymen and cadres appeal to security force shame if or when police activities cause civilian casualties, especially among the very young, old, women, and children. Media handlers or internal insurgent media will be pre-positioned to record footage of any use of government force, and the most sophisticated movements will have cadres in place to shape, package, and distribute messages supportive of the insurgent's goals.

- *Nonviolent civil disturbance.* Organizers should also consider nonviolent civil disturbances as an option for achieving intermediate and long-term objectives. Nonviolent actions can achieve many of the same goals as violent disturbances, and often with longer-lasting effects. These demonstrations can create a spectacle for media attention, raise public awareness of common grievances, disrupt the normal order of civic life, affect the economy, inspire recruitment through a show of strength and persistence, and enable the committed to exercise their revolutionary zeal in a way that builds their confidence and strengthens resolve without exposing them to mass repression. Nonviolent action makes the movement more appealing to those who are sitting on the fence, presents a more sympathetic profile in the media, discourages security-force crackdown while creating an opportunity for the resistance to exploit atrocities, and builds support for the movement by exposing the excessive use of force by government security forces.

Post-Civil Disturbance Phase

3-66. The civil disturbance and its effects should be exploited using traditional local and international media, social media, and cellular and Internet communication mechanisms. The resistance movement should control the release of photographs, interviews, and videos surrounding the event to build internal and external support. After a civil disturbance has subsided, underground elements use a variety of means to exploit the situation. One way of maintaining the interest and emotional involvement of the population is a 24-hour general strike. Workers, especially those in key industries and utilities, are encouraged to protest against the government by staying away from work 24 hours. This is time enough to interrupt vital utilities and affect the entire population. Individuals and property owners are faced with the dilemma of going about their normal routine and facing violence or staying home for a day.

3-67. To turn specific issues into grievances against the government, the underground makes appeals to all workers to join in a united front against the government. Cells in various factories, districts, zones, and businesses demand that their organizations support the strike in the form of a united front. Factory cells incite union members to stay off the job. They seek union sponsorship.

3-68. In order to demonstrate the uncompromising position of the government, the demands against it are usually vague and impossible to meet. Original issues, such as higher wages or repeal of a sales tax, are now changed to antigovernment demands. A call is made for the release of political prisoners, and the police and army are asked to join the rioters. It is customary to insist on nothing less than the complete overthrow of the existing government. These demands can be articulated in protest meetings that keep the public aroused and involved. Committees are formed in every village or city to protest government action. Every attempt is made to get notable and respected citizens to lend their names to the protest.

SABOTAGE

3-69. The terms "sabotage" and "subversion" have distinct military definitions, but in common English usage they are frequently used interchangeably. Sabotage is defined as an act or acts with intent to injure, interfere with, or obstruct the national defense of a country by willfully injuring or destroying, or attempting to injure or destroy, any national defense or war materiel, premises, or utilities, to include human and natural resources. Sabotage is technically a component of subversion because it consists of actions which do contribute to the "undermining of the military, economic, psychological, or political strength or morale of a governing authority." However, subversion generally connotes the actions directed at human beings and meant to undermine the sources of political power, whereas sabotage generally connotes actions directed at physical things and processes and meant to undermine the sources of material power. Nevertheless, there will continue to be instances such as "noncooperation with authorities" which are equally understood as both subversion and sabotage.

GENERAL AND STRATEGIC SABOTAGE CHARACTERISTICS

3-70. Sabotage is an attempt to damage the resources of the government's war effort—military and economic organizations; industrial, food, and commodities production; and public morale and law and order. In most specific instances of general (or "simple") sabotage, the material damage inflicted on the enemy by sabotage is relatively small. However, general sabotage has the effect of cumulatively degrading all forms of enemy strength. Karl von Clausewitz said that war includes "friction" which retards, slows, and frustrates war efforts. Sabotage is like creating man-made friction. General sabotage disrupts normal flows of production, wears everything out faster, and makes everything operate less optimally. The effect of generalized sabotage is to increase enemy expenditure of money, time, and manpower, while hurting his morale and cohesiveness.

3-71. In some cases, however, a single event of strategic sabotage may have significant strategic or operational impact disproportionate to the means employed. A small team of Norwegian underground and special operatives, for example, may have prevented the Nazis from developing the atomic bomb with obvious war-winning ramifications.

Operations GROUSE, FRESHMAN, and GUNNERSIDE

During WWII, the Allies decided to remove the heavy water supply and destroy the Vemork heavy water plant in Telemark, Norway, in order to inhibit the Nazi development of nuclear weapons. Between 1940 and 1944, a sequence of sabotage actions by the Norwegian resistance movement, as well as Allied bombing, ensured the destruction of the plant and the loss of the heavy water produced.

In Operation GROUSE, the British Special Operations Executive (SOE) successfully placed four Norwegian nationals as an advance team in the region of the Hardanger Plateau above the plant. Later in 1942, the unsuccessful Operation FRESHMAN was mounted by British paratroopers. They were to rendezvous with the Norwegians of Operation GROUSE and proceed to Vemork. This attempt failed when the military gliders crashed short of their destination, as did one of the tugs, a Halifax bomber. The other Halifax bomber returned to base, but all the other participants were killed in the crashes or captured, interrogated, and executed by the Gestapo.

> In 1943, a team of SOE-trained Norwegian commandos succeeded in destroying the production facility with a second attempt, Operation GUNNERSIDE. The Norwegian advance team collected intelligence on vital elements of the plant's operation, such as the number, hours of change, and behavioral pattern of the German guards; the layout of the plant; and routes of ingress and egress. The underground made contact with the plant's chief engineer, a Norwegian, who gave them information about the floor plans and the location of critical machinery. In order to plan and rehearse their attack, they built an exact model of the heavy water plant in England.
>
> After nearly a year's planning, an 11-man team infiltrated the area for the actual attack. The team divided into two groups: blocking and demolition. The time for the strike was set for 12:30 a.m. to ensure that the off-duty guards would be asleep and to allow the team 5 hours of darkness to escape. In the actual attack, the blocking unit forced the entrance and covered the German guard barracks, while the demolition unit entered the plant itself through a cable tunnel and set the charge at a predetermined location. Although the explosion was so small it did not arouse the German garrison, it destroyed key apparatus in the plant and 3,000 pounds of heavy water, nearly half a year's production.
>
> These actions were followed by Allied bombing raids. The Germans eventually elected to cease operations and remove the remaining heavy water to Germany. Norwegian resistance forces subsequently sank the ferry *SF Hydro* on Lake Tinnsjo, preventing the heavy water from being removed.
>
> After the war, the SOE judged Operation GUNNERSIDE to be the most successful act of strategic sabotage in all of WWII. The Nazis did not win the race to develop the atomic bomb.

3-72. Sabotage can be low-tech or high-tech. In the Norwegian example, the strategic sabotage mission relied on good intelligence and preparation work by the underground and the daring of the saboteurs. Nevertheless, the explosive charges used and the means of physically emplacing them on the critical equipment were low-tech activities. Today, the ability to surreptitiously emplace malware through the Internet is not only a superlative use of indirect and disproportionate means, it can use cutting-edge technology to achieve strategic sabotage results. Stuxnet, an electronic worm released into the computer networks of the Iranian nuclear enrichment program, is an example of a precision cyber weapon. Stuxnet was designed to target a specific system configuration known to be associated with specialized industrial control processes. Most experts agree that Stuxnet was a product of a state-sponsored computer network operations program; however, the source has remained untraceable. The Stuxnet worm was confirmed by Iranian officials to have affected centrifuges used in the nuclear development process.

3-73. Sabotage can have several purposes in a UW campaign. Planners should obviously seek to degrade the enemy regime or occupying power's resources generally and also identify strategic targets when possible. However, there are other advantages of conducting sabotage:

- Whereas UW is largely about organizing and channeling popular resistance, and because anyone can be trained to conduct simple acts of sabotage, sabotage provides a relatively low-cost and low-risk method for the indigenous populace to participate in resistance.
- Such participation is an "act of war," which means the UW operators are force-multiplying the "combatants."
- Such participation is an easy way to build "esprit de resistance" amongst the population and correspondingly weaken enemy morale and inculcate fear in the security forces. Moreover, sabotage indicates popular resistance to the regime or occupying power in a way difficult to pin on any one individual or group.
- Simple sabotage most often uses natural, ordinary materials and is typically conducted by average citizens with routine access to targets. It is thus an indirect method of "waging warfare" through the populace. By contrast, strategic sabotage may require highly specialized saboteurs

(such as the British commandos in Norway) and/or highly sophisticated techniques or technology (such as computer hacker denial-of-service attacks or the surreptitious insertion of the Stuxnet worm).

- The cumulative effects of general sabotage in a given area can have strategically significant results (for example, the interdiction of railways, roadways, and communications throughout Normandy before and during the Allied landings, or the massive and widespread timed failure of imported computer chips integral to indigenous defense systems prior to a conventional strike). This economy-of-force quality of sabotage is a fundamental characteristic of UW's utility when it is conducted in support of major operations.

TECHNIQUES AND SELECTED CONSIDERATIONS

3-74. Sabotage techniques are only limited by imagination, opportunity, and survivable risk. For obvious safety reasons, a detailed list of specific techniques will not be provided here. However, generalized techniques and selected considerations of employing sabotage follow.

3-75. General or nuisance sabotage is closely related to passive resistance in that it requires neither trained sabotage teams nor carefully selected targets. Sabotage acts in this category usually express individual resistance and take the form of noncooperation, such as deliberate slowdowns on factory production lines, or harassment, such as telephoned bomb threats that force the evacuation and search of buildings and plants.

3-76. Noncooperation sabotage was used extensively in occupied Europe during the German occupation during WWII. Workers slowed their pace of production, went on strike, and refused to help Germans apprehend rebel patriots; postal workers intercepted letters addressed to the Gestapo. The underground pressured doctors into signing medical certificates stating that certain key people were unable to work. During the Cold War, Soviet, Polish, and other Iron Curtain populations used the same techniques against the communist state. Today, strikes by Egyptian air traffic controllers, state workers in Thailand, or food service workers in Greece effectively use coordinated work stoppages to aggravate authorities, burden the economy, and tie down government security forces.

3-77. In WWII, the underground distributed camouflaged pamphlets of sabotage techniques to prompt patriots to sabotage the German occupation. Techniques included methods to slow down production in factories. Sharp metal objects were put in the streets to puncture auto tires. This technique, which virtually halted traffic, was very effective because the only people who had automobiles were German officials and collaborators. Other techniques included failing to lubricate machines IAW maintenance schedules, hiding repair parts, and dropping tools and other foreign objects into moving parts. Today's more complex technology and reliance on electronic and computerized components and digitized transmission provides new government vulnerabilities for the saboteur. Highly sophisticated machinery relies on tighter tolerances that correspondingly make them easier to miscalibrate or break. Ubiquitous electronic components are vulnerable to electromagnetic pulses or power stoppages. The digital commons has opened up unprecedented vulnerabilities to the skilled digital saboteur.

3-78. Where destruction is involved, the traditional and timeless weapons of the citizen-saboteur are salt, nails, candles, pebbles, thread, or any other materials he might normally be expected to possess as a householder or as a worker in his particular occupation; they appear to be innocent. His arsenal is the kitchen shelf, the trash pile, his own usual kit of tools and supplies; they excite no suspicion. The targets of his sabotage are usually objects to which he has normal and inconspicuous access in everyday life. In the 21st century, commonplace items that can be weaponized for sabotage now also include the cell phone, the remote control, the personal computer, and ordinary use of airplanes, trains, and automobiles.

3-79. A different type of simple sabotage requires no destructive tools whatsoever and produces physical damage, if any, by highly indirect means. It is based on universal "human error" opportunities to make faulty decisions, to adopt a noncooperative attitude, and to induce others to follow suit. Making a faulty decision may be simply a matter of placing tools in one spot instead of another. A noncooperative attitude may involve nothing more than creating an unpleasant situation among one's fellow workers, engaging in bickering, or displaying surliness and stupidity. The continuously expanding interconnection of human activities through social media have exponentially enabled and accelerated the potential for such "antisocial" disruption effects, and empowered trained cadres to rapidly mobilize such efforts.

3-80. Motivating the saboteur to begin and continue to conduct sabotage over sustained periods is a special problem. Simple sabotage is often an act which the citizen performs according to his own initiative and inclination; usually, he cannot be closely controlled. The will to conduct sabotage is usually counterintuitive. He must be persuaded that he is acting in self-defense against the enemy or retaliating against the enemy for other acts of destruction. On the other hand, not all potential saboteurs are necessarily motivated by patriotic fervor. Sociopaths and criminal elements can be manipulated or compensated to conduct acts of sabotage in the context of a controlling cadre's operational plan. In addition, one man's patriot can be another man's radical; extremists of all ideologies who are otherwise already willing to use violence are likely to be easier to persuade to expand to sabotage activities.

3-81. The saboteur may have to reverse his thinking and he should be told this in so many words. Where he formerly thought of keeping his tools sharp, he should now let them grow dull; surfaces that formerly were lubricated now should be sanded; normally diligent, he should now be lazy and careless; and so on. Once he is encouraged to think backwards about himself and the objects of his everyday life, the saboteur will see many opportunities in his immediate environment that cannot possibly be seen from a distance. A state of mind should be encouraged that anything can be sabotaged.

3-82. Gains should be stated as specifically as possible for the area addressed: simple sabotage will hasten the day when Commissioner X and his deputies Y and Z will be thrown out, when particularly obnoxious decrees and restrictions will be abolished, when food will arrive, and so on. Abstract verbalizations about personal liberty, freedom of the press, and so on, will probably not be convincing in most parts of the world. In many areas, they may not even be comprehensible. However, there are recent examples which are exceptions to this general rule. Al-Qaeda and other violent Islamist extremists seeking creation of an ideal caliphate, globalist extremists willing to use violence to force a worldwide utopian state, or violent anarchists who want to eliminate the state may be sufficiently motivated by their ideology or dogma to commit acts of sabotage.

3-83. The amount of activity carried on by the saboteur will be governed not only by the number of opportunities he sees, but also by the amount of danger he feels. Bad news travels fast, and simple sabotage will be discouraged if too many simple saboteurs are arrested. The saboteur must learn to commit acts for which large numbers of people could be responsible and for which it would be impossible to be blamed directly. Nevertheless, some deliberately attributable acts can be committed occasionally. For example, dropping a wrench across an electric circuit can be blamed on accident, fatigue, stupidity, ignorance, or other excuses.

3-84. The citizen-saboteur cannot be closely controlled. Nor is it reasonable to expect that simple sabotage can be precisely concentrated on specific types of target according to the requirements of a concrete military situation. Attempts to control simple sabotage according to developing military factors, moreover, might provide the enemy with intelligence of more or less value in anticipating the date and area of notably intensified or notably slackened military activity. On the other hand, when resistance is perceived to be widespread, it can act as a distraction and cover for simultaneous acts of sabotage, as anarchist sabotage cadres may hide within the so-called "Occupy Wall Street" phenomenon, or Muslim Brotherhood agitators can conduct anti-government sabotage within the throngs in Tahrir Square.

3-85. Simple sabotage is more than malicious mischief, and it should always consist of acts whose results will be detrimental to the materials and manpower of the enemy. The saboteur should try to damage only objects and materials known to be in use by the enemy or to be destined for early use by the enemy. It will be safe for him to assume that almost any product of heavy industry is destined for enemy use, and that the most efficient fuels and lubricants also are destined for enemy use. Without special knowledge, however, it would be undesirable for him to attempt destruction of food crops or food products. To the extent that sabotage patterns are programmed by the UW planners or underground, some potential sabotage targets should be spared if the impact on the populace—and resulting popular support for the resistance movement—is too detrimental.

NONCONVENTIONAL ASSISTED RECOVERY

3-86. Personnel recovery is the sum of military, diplomatic, and civil efforts to affect the recovery and reintegration of isolated personnel. Isolated personnel are those U.S. military, DOD civilians, DOD contractor personnel, and others as designated by the President or SecDef who are separated (as an individual or group) from their unit while participating in a U.S.-sponsored military activity or mission and who are, or may be, in a situation where they must survive, evade, resist, or escape. Additionally, when directed by the President or SecDef, personnel recovery support can be provided to other governments, agencies, organizations, and individuals IAW all applicable laws, regulations, and memoranda of agreement or understanding.

3-87. Combatant commanders or their designated JFCs establish joint personnel recovery centers to plan, coordinate, and monitor personnel recovery missions, and to integrate personnel recovery activities with other operations and activities in the assigned operational area. This center is also the JFC's primary coordinator for personnel recovery assistance to other nations or other appropriate civil entities when such assistance is authorized. All U.S. components have some capability to conduct or assist conventional personnel recovery missions.

3-88. SOF provide a unique contribution to a combatant commander's personnel recovery architecture—unconventional assisted recovery, which is nonconventional assisted recovery by SOF. Nonconventional assisted recovery is the recovery of isolated persons by SOF; UW air, ground, and maritime forces; and other government agencies (OGAs) specially trained to develop nonconventional assisted recovery infrastructure and interface with or employ indigenous or surrogate personnel. These forces operate in uncertain or hostile areas where personnel recovery capability is infeasible, is inaccessible, or does not exist, to contact, authenticate, support, move, and exfiltrate isolated personnel back to friendly control. Nonconventional assisted recovery forces generally deploy into their assigned areas before strike operations and provide the JFC with coordinated personnel recovery capability for as long as the forces remain viable. Unconventional assisted recovery terms are defined as follows:

- *Unconventional assisted recovery*: Nonconventional assisted recovery conducted by SOF. Nonconventional assisted recovery is personnel recovery conducted by indigenous/surrogate personnel that are trained, supported, and led by SOF, UW ground and maritime forces, or OGA personnel who have been specifically trained and directed to establish and operate indigenous or surrogate infrastructures. (JP 3-50, *Personnel Recovery*)
- *Unconventional assisted recovery coordination cell*: A compartmented SOF facility established by the joint force special operations component commander, staffed on a continuous basis by supervisory personnel and tactical planners to coordinate, synchronize, and deconflict nonconventional assisted recovery operations within the operational area assigned to the joint force commander. (JP 3-50)
- *Unconventional assisted recovery mechanism*: A recovery mechanism developed and managed by SOF. (Department of Defense Instruction [DODI] 2310.6, *Non-Conventional Assisted Recovery in the Department of Defense*)
- *Unconventional assisted recovery team*: A designated SOF UW ground or maritime force capable of conducting unconventional assisted recovery with indigenous or surrogate forces. (DODI 2310.6)

3-89. The JFC normally designates the joint special operations task force (JSOTF) with overall responsibility for planning, coordinating, and executing all nonconventional assisted recovery operations in support of the personnel recovery plan. The joint force special operations component commander retains operational control of all SOF unconventional assisted recovery forces in the operational area. The joint force special operations component commander exercises control through the operations officer, who designates an unconventional assisted recovery coordination cell director and, when directed, establishes the unconventional assisted recovery coordination cell. The unconventional assisted recovery coordination cell is a compartmented SOF facility staffed on a continuous basis by supervisory personnel and tactical planners who are representative of each nonconventional assisted recovery capability. Nonconventional assisted recovery operations involve the employment of recovery teams and recovery mechanisms.

3-90. As a subset of UW, SOF conduct unconventional assisted recovery unilaterally with indigenous or surrogate personnel or with other governmental agencies, employing compartmented tactics, techniques, and procedures. The military aspects of unconventional assisted recovery require sensitive methods for which SOF are specifically organized, trained, and equipped. In the conduct of unconventional assisted recovery, SOF may be deployed into a JSOA before strike operations to support recovery operations. The intent of unconventional assisted recovery is to bring isolated personnel into contact with, and ultimately into the custody of, a recovery force as soon as possible and then move the isolated personnel to an area where exfiltration to definitive USG control can occur.

Note: IAW 10 USC; DODI 2310.6; USSOCOM Directive 525-21, *(S) Personnel Recovery (U)*; and FM 3-05.220, unconventional assisted recovery is nonconventional assisted recovery conducted by SOF, specifically U.S. Army SF and U.S. Navy sea-air-land teams. Unconventional assisted recovery is a subset and integral component of UW.

3-91. Unconventional assisted recovery is highly sensitive, and such operations may be covert or clandestine. All of the tactics, techniques, and procedures pertinent to Army SF interaction with indigenous forces in UW—particularly those methods which empower insurgent or resistant undergrounds—are useful to unassisted recovery mechanisms. In concert with indigenous partners, procedures and supporting assets are developed to achieve the five specified tasks of nonconventional assisted recovery: contact, authenticate, support, move, and exfiltrate isolated personnel to friendly control. The artful design of these procedures and assets results in the creation of an unconventional assisted recovery mechanism. Further information on unconventional assisted recovery can be found in FM 3-05.220; FM 3-05.231, *Special Forces Personnel Recovery*; FM 3-05.701, *(C) Army Special Operations Forces Resistance and Escape (U)*; and FM 3-05.70, *Survival*.

3-92. Interested readers may also gain insight into the potential range of activities possible in unconventional assisted recovery mechanism design by looking at legacy vignettes. Traditionally, personnel recovery has often been referred to as "escape and evasion" or "escape and recovery." During WWII and the Cold War, for example, military, civilian, and indigenous special operatives set up "E&E networks (or 'nets')" or "E&R nets." There are many historical examples available.

Example Vignettes/Concepts of Escape and Evasion Networks From Previous Eras

Throughout the 20th century, "escape and evasion nets" essentially consisted of escape routes, hideouts or "safe houses," and some secretive organization of shadowy figures to assist the evader through the network in enemy country until he could be delivered to areas under friendly control.

There were three general categories of safe houses: the temporary stopover, the emergency hideout, and the permanent refuge. Couriers and traveling agents used the temporary stopover to facilitate travel. Escapees and persons in danger also used temporary safe houses along escape routes for food, rest, and directions to the next stopover. An operative who became suddenly ill, wounded, or sought by the police could use an emergency hideout. Such safe houses were typically private homes of loyal and reliable persons who were supporters of, but not identified with, the underground movement. Other facilities have also been used as safe houses. Algerian physicians loyal to the Front de Liberation Nationale (FLN) hid evaders as patients in the Algiers Municipal Hospital. The permanent or long-term safe house was often an isolated farm or cabin, a distant encampment, or a location in a nearby nation sympathetic to the underground movement.

> In the 1930s, the Soviet Comintern utilized extensive auxiliary offices and bases for their agents abroad. The Seamen's and Port Worker's International, for instance, controlled seamen's clubs in every major port in the Western Hemisphere. These clubs served as reporting and relocation bases for agents operating in or traveling through the country. Personnel at these auxiliary bases arranged contacts, passports, cover addresses, and funds for agents. When an agent lost contact with his organization, he simply reported to the nearest auxiliary base for food, shelter, funds, and instructions.
>
> In planning riots and demonstrations, the Viet Cong preestablished safe zones in sections of the city where they would store weapons and assemble agitators. They identified shopkeepers and homeowners willing to provide shelter for the demonstrators. The agitators hid in these safe houses until the police had completed their post-demonstration search. Afterwards, the agitators would withdraw from the city by predetermined routes.
>
> The safe houses along an escape-and-evasion network would usually be placed only within one day's travel of each other. The person maintaining the safe house seldom engaged in any other subversive activity that might have drawn attention to him. The underground would have supplied him with extra food, clothes, and any identification papers or documents needed to conduct his portion of the clandestine effort. Each person in the route knew how to reach only the next link, and no one person knew the identity or location of every link. Guides generally escorted the escapee from one link to the next. The guides would meet at a prearranged spot halfway between the two safe houses and neither guide would know the location of the other's safe house.
>
> The Viet Cong infiltration process from North Vietnam to South Vietnam provides an illustration of safe house and fail-safe concepts. After completing training in the North, Communist infiltrators were trucked to the Laotian border just above the demarcation line where they rested for several days before beginning their move southward. An infiltration group usually numbered 40 to 50 men, but once they reached the border they broke up into smaller groups. Each man carried a three- to five-day supply of food, a first-aid packet, hammock, mosquito netting, and similar items. No one was permitted to carry personal papers, letters, or photographs that might be used by the enemy to identify him. The infiltration routes along the Laos-South Vietnam border included way stations. A chain of local guides led the units along a network of secret trails. Each guide knew only his own way station and conveyed troops to the next way station just as the network conveys escapees between safe houses. Conversation was discouraged in transit and only the leader of the group was permitted to speak with the guide. In this manner, the network maintained a degree of security and contained damage if one guide defected or was captured.
>
> During WWII, underground escape-and-evasion nets devised some unusual techniques to pass escapees beyond checkpoints. Police members of the net would handcuff escapees and pass them through the checkpoint as prisoners. Underground members also hid escapees in maternity homes until their passage through the escape route could be secured. In one incident, an escapee was passed through a checkpoint by placing him in an ambulance and having him feign insanity.

GUERRILLA WARFARE

3-93. Guerrilla warfare is the activity most often associated and mistakenly equated with UW. UW is, in fact, a collection of many different activities, most of which are more important than the typically visible and easily understood contribution of organized armed irregulars. Successful UW usually relies much more on political activities, such as effective leadership, insurgent organization building, and broad subversion of regime power, and on effective sustainment of crucial support functions like insurgent logistics and intelligence, than it does on tactical guerrilla warfare victories on the battlefield. The proof of this is the long string of repressed armed rebellions without popular backing throughout all of human history. Those who place all or most of their focus on guerrilla warfare do not understand UW. Guerrilla warfare is only one relatively narrow and limited activity subordinate to UW.

3-94. Nevertheless, while the above statement is true, it is also true that almost all UW endeavors will not succeed without conducting effective guerrilla warfare, the results of which are integrated into a well-conceived UW campaign plan. UW is "warfare" and assumes violence. Some—sometimes much—of such violence is armed propaganda, political murder, and terrorism. Some is the "small warfare" known as guerrilla warfare. Guerrilla terms are defined as follows:

- *Guerrilla*: **An irregular, predominantly indigenous member of a guerrilla force organized similar to military concepts and structure in order to conduct military and paramilitary operations in enemy-held, hostile, or denied territory. Although a guerrilla and guerrilla forces can exist independent of an insurgency, guerrillas normally operate in covert and overt resistance operations of an insurgency.**
- *Guerrilla force*: A group of irregular, predominantly indigenous personnel organized along military lines to conduct military and paramilitary operations in enemy-held, hostile, or denied territory. (JP 3-05)
- *Guerrilla warfare*: Military and paramilitary operations conducted in enemy-held or hostile territory by irregular, predominantly indigenous, forces. (JP 3-05.1)

CHARACTERISTICS OF GUERRILLA WARFARE

3-95. Guerrilla warfare comprises combat operations conducted in enemy-held territory by predominantly indigenous forces on a military or paramilitary basis to reduce the effectiveness, industrial capacity, and morale of the enemy. Guerrilla operations are conducted by relatively small groups employing offensive tactics. In UW as conceived by U.S. Army SF, guerrilla warfare is not an end in itself; it is a component of a larger UW campaign plan which supports the achievement of intermediate objectives en route to an end state political decision.

Nature of Guerrilla Warfare

> *The strategy is one to ten; the tactics are ten to one.*
> Li Tso-peng commentary on Mao's *On Protracted War*

3-96. Guerrillas are irregular soldiers generally operating with firepower, manpower, equipment, and logistical disadvantages compared to the security forces of the regime or occupying power. Guerrillas initially cannot hope to meet and decisively defeat a conventional unit in a pitched battle. Guerrilla units, therefore, must emphasize preservation of their military forces and attack at points most disadvantageous to the enemy. These attacks are normally conducted during periods of low visibility and are directed against isolated outposts, weakly defended locations, or the moving enemy. By recognizing his own limitations and weaknesses, the guerrilla can hope for survival and eventual success. Although typically inferior to the enemy in many ways, the guerrilla is often equal or superior to the enemy in the collection of intelligence information, cover and deception, and the use of surprise to gain and maintain the initiative. The guerrilla exploits these advantages to compensate for his physical disadvantages.

3-97. In guerrilla warfare, the situation is always fluid. Both guerrilla units and enemy counterguerrilla forces move about on the battlefield—the former to avoid decisive engagement and annihilation and to maintain the initiative, and the latter in the attempt to find, fix, and finish the guerrillas. Guerrilla operations wear down and inflict casualties upon the enemy, cause damage to supplies and facilities, and hinder enemy operations. Because guerrilla operations are primarily directed against lines of communications, industrial facilities, and key installations, they impede or interdict the movement of men and materiel and seriously affect the enemy's capability to supply, control, and communicate with his combat forces. In addition, the enemy is compelled to divert manpower and equipment to combat guerrilla activities. The success of guerrilla operations—even the fact that the guerrillas continue to exist—lowers enemy morale and prestige and maintains the morale and will to resist of the indigenous population.

3-98. In the early stages of UW, guerrilla forces are rarely concerned with seizing and holding terrain, primarily because they do not have the strength to do so against government countermeasures. The guerrilla attacks by gaining a momentary advantage of firepower, executes his mission to capture or destroy personnel and equipment, and leaves the scene of action as rapidly as possible. Normally, the guerrilla varies his operations so that no pattern is evident. Guerrilla forces often make use of difficult and inaccessible terrain to aid concealment and frustrate enemy pursuit or occupation. Guerrilla forces will typically be assigned areas of responsibility within which they will operate against government power, constantly move, and gradually develop into insurgent sanctuary enclaves relatively free from enemy interference. Ultimately, these enclaves will grow in size and autonomy and will be governed by the insurgents as liberated territory. In the most successful examples of UW, the guerrillas eventually operate in close coordination and more openly with insurgent regulars or outside supporters' conventional maneuver units.

Types of Guerrilla Warfare Operational Missions

3-99. Guerrilla warfare missions are generally of three basic types. Guerrilla warfare may be conducted as part of a standalone UW initiative where the guerrilla force is the predominant or only friendly armed force in the AO. Alternatively, forces conducting guerrilla warfare may be a small adjunct to a conventional force main effort when UW is only one line of operation in major combat operations. Thirdly, in some cases, guerrilla warfare may continue after linkup with friendly forces has occurred.

Guerrilla Warfare Missions in Support of the Theater Commander's Unconventional Warfare Campaign

3-100. UW may represent the entire U.S. strategic initiative or response to an opponent. In these cases, U.S. overt support will be limited. U.S. large-scale intervention is not anticipated and the main effort is the UW campaign itself. U.S. operations will be mostly limited to low-visibility, clandestine, and covert activities. Indigenous emphasis is on the tactical techniques of guerrilla warfare themselves and their operational usefulness to the UW campaign plan. These operations assume the least amount of U.S. direct control. U.S. Army SF and other U.S. operatives typically provide combat advice and assistance.

Guerrilla Warfare Missions to Assist Conventional Forces Engaged in Combat Operations

3-101. UW may represent a line of operation in support of large-scale U.S. intervention involving major operations. Although the indigenous tactical techniques of guerrilla warfare remain the same, the emphasis in this case is their operational usefulness to conventional major operations. These operations assume a greater degree of U.S. direct control, possibly with U.S. tactical commanders exercising operational control of guerrilla forces. U.S. Army SF also function as a liaison for the tactical commander's more direct authority.

Guerrilla Force Missions Conducted After Linkup With Friendly Forces

3-102. Missions may be assigned to guerrilla forces after linkup with friendly conventional forces. If combat operations continue against a major opponent still in control of denied territory, such missions may constitute continuing UW in new areas of operation. In such a case, U.S. Army SF and other U.S. operatives are very likely to continue providing combat advice, assistance, and liaison. If combat against major opponents has culminated in U.S. or coalition victory, such missions may represent some stage of Phase VII (Transition) whereby the indigenous guerrilla forces' status has changed to an adjunct of the new governing authority's security posture—something more like "light infantry," "scouts," or "constabulary." This would probably represent an end to UW operations, and subsequent activities would likely represent support to a foreign internal defense mission in support of the new government. The authorities of the new host-nation government, the degree of any direct U.S. operational control, and the continued involvement by U.S. Army SF or other U.S. operatives would then vary by specific circumstances.

GUERRILLA WARFARE MISSIONS IN SUPPORT OF THE THEATER COMMANDER'S UNCONVENTIONAL WARFARE CAMPAIGN

3-103. Guerrilla forces conduct missions and activities supportive of the UW campaign plan. These missions will have a range of tactical, operational, and strategic objectives; will have both immediate and long-term effects on the enemy; and will have direct military and indirect political and psychological implications. These missions and activities comprise the basic techniques of guerrilla warfare. They consist of the following:

- Offensive guerrilla combat operations are comprised of the raid and ambush. These techniques are the fundamental tactics of all guerrilla operations throughout the history of warfare.
- Interdiction is a major operational technique and emphasis in UW and is therefore a basic component of guerrilla warfare. Interdiction—properly planned and employed as part of a comprehensive UW campaign plan—is probably the single most important guerrilla warfare activity that guerrilla forces conduct.
- PSYACTs may be executed by UW indigenous partners who are in turn supported by U.S. MISO units that advise and assist in the development and employment of guerrilla information capabilities. All operations have a psychological impact and all should be conducted in a manner that will enhance organization and favorable influence of designated target audiences. The psychological impact of guerrilla operations may, in fact, be more significant—individually or cumulatively—than the material combat results. Therefore, guerrilla force activities affect influence efforts. Nevertheless, influence activities are part of the larger context of the UW campaign plan and will be managed primarily by elements other than the guerrilla forces themselves.
- Intelligence tasks are a basic soldier responsibility. Guerrilla forces will conduct reconnaissance and surveillance missions as they would other combat patrols and will report observed tactical information as a matter of standard operating procedures. Like psychological operations, indigenous conduct of specific and special intelligence activities will be conducted primarily by members of the underground and auxiliary.
- Evasion-and-escape mechanisms are developed to assist in the recovery of separated friendly personnel. Although guerrilla units assist evasion-and-escape activities, such operations are likewise conducted primarily by the underground and auxiliary.
- Defensive combat operations are almost a misnomer because the overwhelming majority of guerrilla force defensive activities are passive. Especially in the early periods of a resistance or insurgency, guerrilla forces are at a marked strength disadvantage compared to government or occupying power security forces. In most cases, guerrilla forces forced to defend themselves from security force attack connotes loss of security and loss of the initiative. In most such cases, the guerrilla force is likely to be destroyed. Therefore, guerrillas defend themselves primarily through low-visibility or high-mobility elusiveness.

Note: UW activities are broader than those typically conducted primarily by guerrilla forces and include such activities as PE, subversion, and sabotage, which are primarily conducted by the underground. Moreover, although guerrilla forces do conduct indigenous information activities (as part of subversive techniques), do support evasion as part of unconventional assisted recovery, and do contribute to resistance intelligence operations, these activities are discussed separately from guerrilla warfare.

Offensive Combat Operations

3-104. Raids and ambushes are the principal offensive techniques of the guerrilla force. Raids and ambushes may be combined with other actions, such as mining and sniping, or these latter actions may be conducted independently. All such tactical actions can be integrated into operational plans to interdict enemy lines of communications, key areas, military installations, and industrial facilities. Operational plans will usually drive the selection of tactical objectives. Operational plans must routinely include consideration of second- and third-order effects of any combat actions on both the enemy and the local population.

Chapter 3

3-105. Careful and detailed planning is a prerequisite for effective guerrilla combat operations. Depending on the resistance movement, the area command will typically issue general operational guidance and dispersed units will use maximum tactical latitude—often because there is no alternative—to achieve the command's objectives. Prior to initiating combat operations, a detailed intelligence collection effort is made in the projected objective area and where possible the target or objective area will be kept under surveillance up to the time of attack. While the resistance may be weak overall strategically, the individual guerrilla force units must choose tactical targets that they can seal off, overwhelm, and dominate for a short specific period. That same guerrilla force's relative weakness mandates an emphasis on tactical surprise on initiation and rapid preplanned withdrawal along alternate routes when concluded.

3-106. Tactical objective selection is very much a function of guerrilla force capabilities and limitations. Guerrilla forces by definition are relatively weak and under-equipped for communications and mobility. If guerrilla forces are widely dispersed and are to converge on an objective, yet have inadequate means of communication, unforeseen contingencies and reactive responses must be carefully considered. Likewise, if lack of mobility would put withdrawing guerrilla forces at a disadvantage to pursuing forces, countervailing measures must be implemented. Engagement by superior enemy reaction forces could result in unacceptable risk to force. These represent serious concerns that supporting SF teams can help mitigate.

3-107. Tactical objective selection and the tactical concept of the operation will also be a function of desirability as assessed by the criticality, accessibility, recuperability, vulnerability, effect, and recognizability (CARVER) matrix. Once targets are selected, control measures such as zones of action, axes of advance, and limits of advance, and supporting efforts such as mission support sites are established. (A *mission support site* is a **preselected area used as a temporary base or stopover point. The mission support site is used to increase the operational range within the joint special operations area.**) Premission training and rehearsals are conducted. SF Soldiers are particularly effective in advising and assisting with such planning, training, and execution considerations.

Raids

3-108. A *raid* is an operation to temporarily seize an area in order to secure information, confuse an adversary, capture personnel or equipment, or to destroy a capability culminating with a planned withdrawal (JP 3-0). Raids are conducted by guerrilla units to destroy or damage supplies, equipment, or installations such as command posts, communication facilities, depots, radar sites, and so on; capture supplies, equipment, and key personnel; or cause casualties among the enemy and his supporters. Other effects of raids are to draw attention away from other operations, keep the enemy off balance, and force him to deploy additional units to protect his rear areas. Raids—like all other combat operations—can also be used as confidence targets and training events for the guerrilla raid force itself.

3-109. The size of the raid force depends upon the mission, nature, and location of the target and the enemy situation. The raid force may vary from a single squad attacking a police checkpoint or unprotected rail lines, to a multi-battalion raid attacking a large garrison or supply depot. Regardless of size, the raid force consists of three basic elements: assault, support, and security. The assault element is organized and trained to accomplish the objectives of the raid. It consists of a main action group to execute the major task of the raid mission (such as blow up a transformer, capture a key individual, or assault a garrison) and may include personnel detailed to execute special tasks (such as place explosives, handle hostages, or breach perimeters). The support element supports the assault element by concentrating high-volume direct and any indirect fires. The security element supports the raid by preventing the enemy from reinforcing or escaping. Additionally, the security element covers the withdrawal of the assault element and acts as a rear guard for the raid force. The size and positioning of the security element depends upon the enemy's capability to intervene in or react to the operation.

3-110. Movement to and withdrawal from the objective are key raid considerations and an intensive intelligence effort precedes each operation. Insurgent intelligence cells, auxiliary members, and guerrilla force reconnaissance elements conduct reconnaissance of the routes to the target and, if possible, of the target itself. Local auxiliary sources may be required to furnish guides, transports, or mission support sites. Surveillance of the target is continuous up to the time of the attack. The raid force commander exercises extreme caution to deny the enemy any indications of the impending operation.

Activities

3-111. Movement to the objective area is planned and conducted to allow the raid force to approach the target undetected. Movement may be over single or multiple routes. The preselected route or routes terminates in or near one or more mission support sites. During movement, every effort is made to avoid contact with the enemy. Upon reaching the mission support site, security groups are deployed and final coordination takes place prior to movement to the attack position (Figure 3-1).

Figure 3-1. Movement to an objective

3-112. Withdrawal is accomplished in a manner designed to achieve maximum deception to the enemy and to facilitate further action by the raid force. The various elements of the raiding force withdraw, in order, over predetermined routes through a series of rallying points. Frequently, the raid force disperses into smaller units, withdraws in different directions, and reassembles at a later time and at a predesignated place to conduct other operations. Elements of the raid force may conduct further operations, such as an ambush of the pursuing enemy force, during the withdrawal.

3-113. Should the enemy organize a close pursuit of the main body, the security element assists by fire and movement, distracting the enemy and slowing him down. Elements of the raiding force that are closely pursued by the enemy do not attempt to reach the initial rallying point, but on their own initiative lead the enemy away from the remainder of the force and attempt to lose him by evasive action over difficult terrain. If the situation permits, an attempt is made to reestablish contact with the raid force at other rallying points or to continue to the base area as a separate group. When necessary, the raiding force separates into small groups or even individuals to evade close pursuit by the enemy (Figure 3-2, page 3-26).

Chapter 3

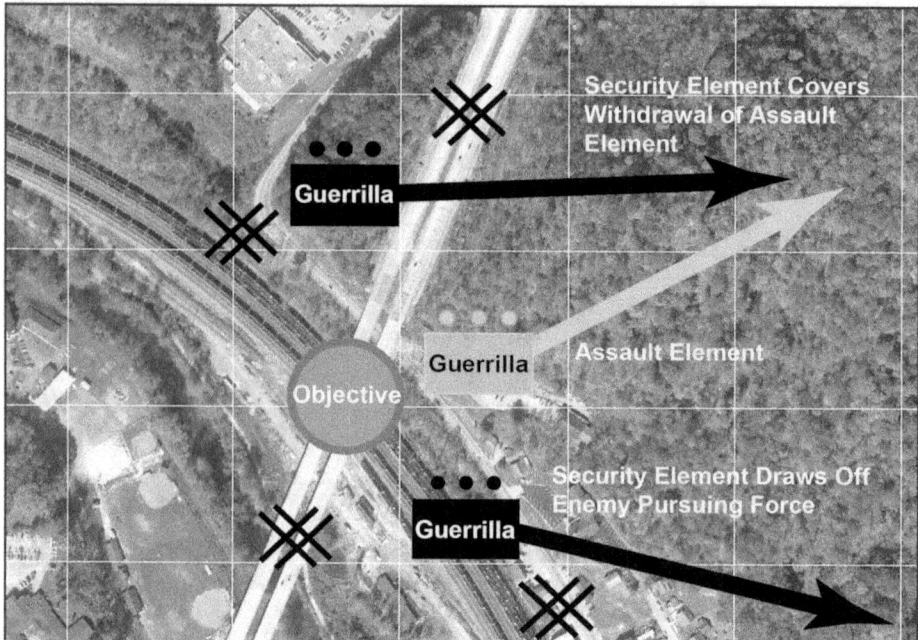

Figure 3-2. Withdrawal from an objective

Ambushes

3-114. An ambush is a form of attack by fire or other destructive means from concealed positions on a moving or temporarily halted enemy. In an ambush, the enemy sets the time and the attacker, the place. Ambushes are conducted to destroy or capture personnel and supplies; harass and demoralize the enemy; delay or block movement of personnel and supplies; and canalize enemy movement by making certain routes useless for traffic. The result usually is concentration of the majority of movements to principal roads and railroads where targets are more vulnerable to attack by other theater forces. As in all guerrilla force combat operations, the area command will issue general operational objectives, which may then be best achieved by selected tactical ambush operations.

3-115. Like the raid force, the ambush force is organized into assault, support, and security elements. The assault element conducts the main attack against the ambush target, which includes halting the column, killing or capturing personnel, recovering supplies and equipment, and destroying unwanted vehicles or supplies that cannot be moved. The support element supports the assault element by concentrating high-volume direct and any indirect fires. The security force isolates the ambush site using fires, demolitions, or roadblocks.

3-116. Preparation for an ambush is similar to that of a raid except that selection of the ambush site is an additional key consideration. Planning factors include the following:
- Will the operation be a single ambush against one column or an area ambush, or a series of ambushes against one or more routes of communication? If it will be a series of ambushes, how are they to be sequenced to achieve maximum effectiveness?
- What are the probable sizes, strength, and composition of the enemy force that is to be ambushed, formations likely to be used, and enemy reinforcement capability and reaction times?

- Finding terrain along the route favorable for an ambush is a key consideration; natural conditions should provide concealment, physical security, and enhance attacker destructive effectiveness and channel and contain the target into the kill zone.
- Terrain considerations include finding unobserved routes of approach and withdrawal to and from the objective.
- Timing of the ambush should provide maximum advantage to the attacker and maximum disadvantage to the target. Most ambushes are conducted during periods of limited visibility.

3-117. Intelligence preparation will be similar to those for a raid with one key exception—effort must be expended to establish enemy traffic patterns and strengths. However, analysis and prioritization of these patterns should already be a central part of operational targeting already. Like the use of raids, the value of ambushes is not limited only to their tactical results; their primary function is to contribute to the imposition of operational effects on the enemy. Movement to and from the objective, training and rehearsals, mission support site preparations, and security measures will be essentially the same as for raids. Given enough damage consistently, the enemy may even abandon a route—and access to the terrain it runs into—entirely. This can be a huge operational achievement providing an additional "liberated" area, as well as a potentially strategic psychological victory.

3-118. There are a few special ambush situations to consider. Most ambushes are conducted relatively close to the enemy kill zone and typically involve members of the assault element briefly occupying the kill zone to capture prisoners or to capture and destroy material—a "near ambush." However, if the purpose of an ambush is not to destroy but merely to harass, the ambush site selection and engagement methods will have to be modified. Such "far ambushes" have the insurgent advantage of putting more distance and possible terrain features between the ambush force and enemy forces.

3-119. Ambushes against columns protected by armored vehicles depend upon the type and location of armored vehicles in a column and the weapons of the ambush force. If possible, armored vehicles are destroyed or disabled by fire of antitank weapons, land mines, improvised explosive devices, Molotov cocktails, or by throwing hand grenades into open hatches. An effort is made to immobilize armored vehicles at a point where they are unable to give protection to the rest of the convoy and block the route of other supporting vehicles.

3-120. Ambushes against moving railroad trains may be subjected to harassing fire, but the most effective ambush involves derailing the train. The locomotive should be derailed on a down grade, at a sharp curve, or on a high bridge. This causes most of the cars to overturn and results in extensive casualties among passengers. It is desirable to derail trains so that the wreckage remains on the tracks to delay traffic for longer periods of time. Fire is directed on the exits of overturned coaches and designated groups armed with automatic weapons rush forward to assault any coaches which are still standing. Other groups take supplies from freight cars and then set fire to the train. Rails are removed from the track at some distance from the ambush site in each direction to delay the arrival of reinforcements by train. In planning the ambush of a train, guerrilla forces must remember that the enemy may include armored railroad cars in the train for its protection and that important trains may be preceded by advance guard locomotives or inspection cars to check the track.

3-121. Waterway traffic like barges, ships, and other craft may be ambushed in a manner similar to a vehicular column. The ambush party may be able to mine the waterway and thus stop traffic. If mining is not feasible, fire delivered by direct-fire heavy weapons can damage or sink the craft. Fire should be directed at engine room spaces, the waterline, and the bridge. Recovery of supplies may be possible if the craft is beached on the banks of the waterway or grounded in shallow water.

Interdiction

3-122. UW forces use interdiction as the primary means of accomplishing operational objectives. Interdiction is designed to prevent or hinder, by any means, enemy use of an area or route. Interdiction is the cumulative effect of numerous smaller offensive operations, such as raids, ambushes, mining, and sniping. Enemy areas or routes that offer the most vulnerable and lucrative targets for interdiction are industrial facilities, military installations, and lines of communication. The results of planned interdiction programs include the following:

- Effectively interfering with the movement of enemy personnel, supplies, equipment, and raw material.
- Destroying storage and production facilities.
- Destroying military installations.
- Frustrating enemy ability to accurately locate guerrilla bases by analyzing guerrilla operations.
- Causing the enemy to overestimate the strength and support of the guerrilla force.
- Demoralizing the enemy, lessening his will to fight, and modifying his movement, access, support, and control patterns.

3-123. Interdiction targets should usually be selected based on their achievement of operational objectives supporting the overall UW campaign plan. Therefore, suitable targets for interdiction are typically facilities and material utilized by the regime or occupying power to maintain political power and enable security forces. Major interdiction target systems include—

- *Transportation.* This includes railroads, highways, waterways, airways, to include physical networks, controlling facilities, and automotive machines and stock.
- *Communication.* This includes telephone, telegraph, radio, television, and computer production and broadcast infrastructure; public address systems and venues; newspaper, magazine, and book production and distribution networks.
- *Industry.* This includes anything supporting government or occupier power, but especially manufacturing facilities for weapons, aircraft, vehicles, ammunition, and shipping, to include support and distribution infrastructure.
- *Power.* This includes electric, nuclear, chemical, hydro, wind, solar, and so on generation and processing facilities and distribution networks.
- *Fuel.* This includes gas, oil, diesel, and so on generation and processing facilities and distribution networks.
- *Military.* This includes military installations and personnel and all relevant supporting infrastructure.
- *Human nodes.* This includes key human nodes in the formal and informal state civilian power structures and key popular or nonstate influence brokers or collaborators.

Note: Raids and ambushes are critical components of interdiction campaigns without being subordinate components of interdiction. Although UW planners should always put tactical actions into the context of operational and strategic plans—and raids and ambushes are common techniques contributing to campaign interdiction plans—raids and ambushes are sometimes justified by themselves. For example, a strategic raid or ambush that results in the killing or capture of a strategic enemy leader, the destruction of weapons of mass destruction (WMD) in enemy hands, or the liberation of large numbers of friendly personnel from enemy captivity are valuable combat operations regardless of any broader interdiction implications.

3-124. There are a few special considerations when planning interdiction campaigns:

- Interdiction is both an action (operation) and an effect created by a combination of insurgent or resistance actions. Commanders and UW planners should remember that all methods can contribute to an interdiction plan, including acts of sabotage and subversion by elements of the underground and auxiliary.
- Actions should be directed against the primary and alternate critical elements of each target system; the quality of political, military, economic, social, information, infrastructure, physical

environment, and time (PMESII-PT), CARVER, and other methods of target analysis is important.
- Commander's intent is particularly important in interdiction operations because effects achieved are more important in interdiction than literal tactical results achieved. For example, destruction of a bridge (literal interdiction result) to interdict stream crossings (commander's intent) does little good if a usable ford is nearby.

3-125. While much of target analysis is concerned with selecting targets to destroy or incapacitate, another critical function is to identify which targets are irreplaceable or would require too much investment of time and resources to reconstruct either by or for the new government. If an infrastructure target is identified as necessary to disrupt, primacy must be given to determining what part of the system can be most easily destroyed or removed and then quickly replaced by friendly forces/entities. This "reverse CARVER" principle is based on the fundamental logic that UW ties much of its success to the populace support and governmental function within the UW operational area. Although destruction of key civil infrastructure can immediately achieve desired effects, it ultimately sets conditions for long-term difficulty or failure.

Mining (Including the Use of Improvised Explosive Devices) and Sniping

3-126. Mining affords the area commander a means of interdicting enemy routes of communication and key areas with little expenditure of manpower. Mines allow the user to move away from the mined site before the enemy activates them. The planned use of mines as an interdiction technique also has a demoralizing effect on enemy morale. Mines may be employed in conjunction with other operations, such as raids, ambushes and sniping, or used alone. When used alone they are emplaced along routes of communication or known enemy approaches within an area at a time when traffic is light. This allows personnel emplacing the mines to complete the task without undue interference and then make good their escape. Mines can be used to cover the withdrawal of a raiding or ambush force to slow enemy pursuit. Mines can also be emplaced around enemy installations to cause casualties to sentinels and patrols.

3-127. Sniping is another interdiction technique. It is economical in the use of personnel and has a demoralizing effect on enemy forces. A few trained snipers can cause casualties among enemy personnel, deny or hinder use of certain routes, and require the enemy to employ a disproportionate number of troops to drive off the snipers. Snipers may operate to cover a mined area, as part of a raiding or ambush force, or by themselves. Snipers operate best in teams of two, alternating the duties of observer and sniper.

Railroad Systems

3-128. Railroads present one of the most profitable and easily accessible target systems for attack by guerrilla forces. Most nations rely heavily upon railroad traffic for movement of heavy military equipment. Therefore, guerrilla interdiction attacks against rails can have far-reaching operational and strategic effects. Railroads are characterized by open stretches of track and vulnerable switches, repair facilities, power plants and substations, and coal and water supplies providing unlimited opportunities for attack. Rail crews are subject to interdiction through intimidation.

3-129. Railroad tracks are easily attacked by guerrilla units because it is almost impossible to guard long stretches of track effectively. Lightly armed, mobile guerrilla units can inflict heavy damage on tracks. Attacks on open tracks use fewer explosives than attacks on other railroad installations and components. A small guerrilla unit can destroy a considerable amount of railroad track in one night. It is possible for a small group of guerrillas working regularly to keep a single track out of operation permanently. Attacks on tracks should cover a wide area. Telegraph and telephone lines along the railroad are cut simultaneously.

3-130. Critical equipment and facilities may be harder to attack because they are usually guarded. However, careful mission planning, CARVER analysis, and skilled execution can still make repair facilities, reserve stocks of equipment, railroad cranes, and other critical items vulnerable. Often the underground or auxiliary will have better access to rail facilities than will guerrillas.

3-131. Rolling stock may be simultaneously attacked with track interdiction. Demolition of tracks at the time when trains are passing can increase the damage to the tracks and track bed, result in captured supplies, kill and wound enemy personnel, or liberate prisoners. Bridges, tunnels, and narrow railway passes, although usually well-guarded, are critical points for train attacks. Trains moving through areas menaced by guerrillas move slowly and are guarded. Attacks on guarded trains require well-trained and well-armed guerrillas. Heavy weapons and explosives are useful in attacking guarded trains.

3-132. Early interdiction of railroads after hostile occupation may interfere with the enemy's offensive momentum and may forestall large-scale deportation of civilian populations. Cumulatively, the primary effect of interdiction of railroads is disruption of the enemy's flow of supplies, movement of troops, and industrial production. Disruption of orderly processes may result in the accumulation of sizable targets jammed up at rail terminals, junctions, and marshalling yards. These targets then make lucrative targets susceptible to attack by other service components. Interdiction will increasingly deplete reserves of repair materials, will increase the burden upon enemy security forces and repair crews, and may force transfer to more costly and ponderous highway traffic.

Highway Systems

3-133. Highways are less vulnerable targets than railroads. Damage inflicted is more easily repaired and repairs require fewer critical materials and less skilled labor. Bridges, underpasses, tunnels, cuts, and culverts are vulnerable points on road networks. Sections of road that may be destroyed by flooding from adjacent rivers, canals, or lakes are also vulnerable. In addition, a road may be interdicted by causing rock slides or landslides.

3-134. Since highways have fewer vulnerable spots, it is likely that these points will be heavily defended. If so, it is better to concentrate on attacking enemy convoys and columns using the highways. In the initial stages of hostilities, small bridges, tunnels, and levees may be insufficiently protected. As guerrilla attacks increase in frequency and effect, enemy security forces increase protection of these likely guerrilla targets.

3-135. Where the roads cannot be destroyed, traffic is interrupted by real and dummy mines. Ambushes are conducted when suitable terrain is available. Long-range fires from positions away from roads disrupt enemy traffic. Points for interdiction are selected in areas where the enemy cannot easily reestablish movement by making a short detour.

Waterway Systems

3-136. Like railroads, waterways typically carry heavy and big-bulk cargos, the interdiction of which could seriously impact the enemy's economy and military sustainability. The most vulnerable portions of waterway systems are electrical installations, dams, and locks which are usually well guarded. The destruction of these installations can disrupt traffic effectively for long periods. Although they are more recuperable, signal lights, beacons, and channel markers can be effectively attacked. Sinking vessels in restricted channels by floating mines, limpets, or fire from heavy-caliber weapons may be effective in blocking waterway traffic. Dropping bridges into the waterway, creating slides where the channel is narrow, and destroying levees all hinder ship movement on waterways. Personnel who operate the waterway facilities, such as pilots and lock operators, may be eliminated. These personnel are not easily replaced and their loss will affect operation of the waterway. Mines and demolitions charges may be placed at strategic points on the waterway.

Airways Systems

3-137. Airways are interdicted by attacking those facilities that support air movement. Air terminals; communications systems; navigational systems; petroleum, oils, and lubricants (POL) dumps; maintenance facilities; and key personnel are targets for attack. Since air traffic is dependent upon fuel, lubricants, spare parts, and maintenance tools, lines of communications and installations providing these items are attacked. The specific conditions of the operational environmental will indicate the criticality of the regime's or occupying power's reliance on airways.

Activities

Communication Systems

3-138. Wire and fiber optic cable communications, wireless devices and transponders, and ground-based components of satellite relay systems are all potentially vulnerable to guerrilla attack. However, destruction of a single axis of a wire or cable or the destruction of a single relay tower or ground installation will seldom result in the complete loss of adversary communications. Alternate routing is normally available and such rerouting often occurs automatically and almost instantaneously. The degree of such communication system redundancy will vary depending on the technology level in the JSOA. However, even in systems with significant redundancy, the destruction of any portion of the system tends to overload the remaining facilities thereby degrading the effectiveness of the adversary's system and taxing his repair and maintenance assets. Moreover, multiple simultaneous attacks, carefully planned and sequenced series of attacks, or persistent cumulative series of attacks may be able to stop communications completely, even if only temporarily or in a small targeted area.

3-139. Cables and wires can be cut and wire poles can be felled. Relay towers and masts can be brought down or the satellite transceivers can be destroyed. Underground cable often runs through concrete conduits and requires more time to destroy. Repair of cables can be delayed by removing a section of the cable. Destruction of physical Internet server sites, telephone central offices and repeater stations, or the supporting power facilities causes greater damage and takes longer to repair than cutting cables or dropping towers. However, radio, television stations, and Internet server sites may be located in well-protected areas and difficult to attack. Destruction of the relay towers and masts, antenna sites, and/or the transmission lines and cables is usually easier to accomplish than destruction of the receiver or transmitter station. Finally, high-technology developments may allow a virtual interdiction of adversary communication systems. However, this is more likely to be done by the underground or support elements in third-party countries than by the guerrilla force itself.

Power Systems

3-140. Power lines are vulnerable to attack much in the same manner as wire communications. Large transmission towers often require demolitions for destruction. Critical points in any power system are the transformer stations. If these stations are not accessible to attacks by guerrilla units, long-range fire from small- or large-caliber weapons may disrupt their operations. Power-producing plants and steam-generating plants may be too heavily guarded for raid operations. To disable them, guerrilla forces should concentrate on interdicting the fuel supply or other support infrastructure.

Water Supply Systems

3-141. The disruption of water lines supplying industries can often be profitably accomplished; water supplies generally are conducted through underground pipelines, and may be destroyed with explosive charges. Raids against reservoir facilities and purification plants are also feasible, but the possible effects upon the civilian population must be considered.

Fuel Supply Systems

3-142. Petroleum and natural gases for an industrial area are usually supplied by pipelines; damage to lines inflicted by rupture and ignition of fuel is considerably greater than damage inflicted on water lines. Large storage tanks at either end of a pipeline are highly vulnerable to weapons fire, especially when using incendiary projectiles. Contaminating agents may be injected into pipelines or fuel tanks.

Defensive Operations

3-143. Guerrilla operations are primarily offensive in nature. Guerrilla units with their relatively light weapons and equipment are normally inferior in strength and firepower to organized enemy forces. They should not, therefore, undertake defensive operations unless forced to do so or in support of special operations conducted by other theater forces. When the enemy attacks, guerrillas defend themselves by movement and dispersion, by withdrawals, or by creating diversions. Whenever possible, defensive operations are accompanied by offensive actions against the enemy's flanks and rear.

Chapter 3

Passive Measures

3-144. Most guerrilla defense is passive. Guerrilla commanders typically make use of distant, inhospitable, and rugged terrain to impede the enemy's ability to find, fix, and pursue them. Guerrilla forces rely heavily on intelligence measures that provide advance warning of impending large-scale counterguerrilla operations. For this reason, coordination with auxiliary inner and outer security rings and underground intelligence cells is a critical early and ongoing requirement in UW. Selected conditions or indicators that might indicate impending enemy offensives in guerrilla operational areas include—

- Advent of suitable weather for extensive field operations.
- Arrival of new enemy commanders.
- Any change in the conventional battle situation which releases additional troops for counterguerrilla operations. Such changes include enemy victories over allied conventional forces, a lull in active operations, and a reduction in the size of the battle area.
- Increase in the size of local garrisons or the arrival of new units in the area, especially if these are combat troops or troops with special counterguerrilla capabilities, such as radio direction finding units; chemical, biological, and radiological units; rotary-winged aircraft; or mountain, airborne, or reconnaissance troops.
- Extension of enemy outposts, increased patrolling, and aerial reconnaissance.
- Increased enemy intelligence effort against the guerrillas.

3-145. Upon receiving information that indicates the enemy is planning a counterguerrilla campaign, the commander should increase his own intelligence effort, determine the disposition and preparedness of his subordinate units, and review plans to meet the anticipated enemy action. Guerrilla forces must always be prepared to displace.

Active Measures

3-146. To divert the enemy's attention, the commander may have diversionary activities initiated in other areas. Likewise, he may intensify his operations against enemy lines of communications and installations. Full utilization of underground and auxiliary capabilities assists diversionary measures. In preparing to meet enemy offensive action, key installations within a guerrilla base are moved to an alternate base and essential records and supplies are transferred to new locations while those less essential are destroyed or cached in dispersed locations. In the event that the commander receives positive intelligence about the enemy's plans for a major counterguerrilla operation, he may decide to evacuate his bases without delay.

3-147. The commander may decide to delay and harass the advancing enemy. The object is to make the attack so expensive that the enemy will terminate operations. First, security activities on the periphery as well as within any bases are accelerated. Maximum utilization is made of the defensive characteristics of the terrain, ambushes are positioned to inflict maximum casualties and delay, and antipersonnel mines are employed extensively to harass the enemy. As the enemy overruns various strongpoints, the defenders withdraw to successive defensive positions to delay and harass again. When the situation permits, they may disperse, pass through the line of encirclement, and initiate attacks on the enemy's flanks, rear, and supply lines. If the enemy is determined to continue his offensive, the guerrilla forces should disengage and evacuate the area, dispersing into small elements that will rally elsewhere at predetermined locations. Under no circumstances does the guerrilla force allow itself to become so engaged that it loses its freedom of action and permits enemy forces to encircle and destroy it.

3-148. A counterinsurgent's encircling maneuver is the greatest danger to guerrilla forces because it prevents the guerrillas from maneuvering. Once the counterinsurgent has succeeded in encircling a guerrilla force, he may adopt one of several possible courses of action (COAs). The simplest is to have his troops close in from all sides, forcing the guerrillas back until they are trapped in a small area which is then assaulted. Differences in terrain make it almost impossible for counterinsurgent troops to advance at an equal rate all around the perimeter, thus creating the possibility of gaps between individuals and units. In other cases, the counterinsurgent may decide to break down the original circle into a number of pockets which will then be cleared one by one. In this situation, the guerrillas must either break out or escape through gaps which may appear as counterinsurgent forces are maneuvering into new positions.

Activities

3-149. Perhaps the most difficult situation for guerrillas to counter with is an assault after encirclement has been accomplished. In this maneuver, counterinsurgent forces on one side of the encircled area either dig in or use natural obstacles to block all possible escape routes, while other counterinsurgent forces on the opposite side of the encirclement advance driving the guerrillas against the fixed positions. As the advance continues, counterinsurgent forces which were on the remaining two sides are formed into mobile reserves to deal with any breakouts.

3-150. A guerrilla commander must be constantly on the alert for indications of encirclement. When he receives indications that an encircling movement is in progress, such as the appearance of enemy forces from two or three directions, the guerrilla commander immediately maneuvers his forces to escape while enemy lines are still thin and spread out, and coordination between advancing units is not yet well established. Records and surplus equipment are either cached or destroyed. Thus, the guerrilla force either escapes the encirclement or places itself in a more favorable position to meet it. If, for some reason, escape is not initially accomplished, movement to a ridgeline is recommended. The ridgeline affords observation, commanding ground, and allows movement in several directions. The guerrillas wait on this high ground until periods of low visibility or other favorable opportunity for a breakthrough attempt occurs. If gaps between enemy units exist, combat detachments seize and hold the flanks of the escape route. When no gaps exist in enemy lines, these detachments attack to create and protect an escape channel. The breakthrough is timed to occur during periods of poor visibility, free from enemy observation and accurate fire. The breakout can be aided by guerrilla diversionary attacks at other points of the encircling perimeter.

3-151. If the breakout is successful, the guerrilla force should increase the tempo of its operations whenever possible, thus raising guerrilla morale and making the enemy cautious in the future about leaving his bases to attack the guerrilla areas. If the breakout attempt is unsuccessful, the commander divides his force into small groups and instructs them to infiltrate through the enemy lines at night or hide in the area until the enemy leaves. This action should be taken only as a last resort, as it means the force will be inoperative for a period of time and the morale of the unit may be adversely affected. Reassembly instructions are announced before the groups disperse.

GUERRILLA WARFARE MISSIONS TO ASSIST CONVENTIONAL FORCES ENGAGED IN COMBAT OPERATIONS

3-152. When the JSOA exists within and in support of a joint operational area, operational control of the guerrilla forces concerned may be passed to the JTF commander. This happens when UW represents a line of operation in support of large-scale U.S. intervention involving major operations. The classic example of UW—WWII in both European and Pacific Theaters—represents missions in support of a larger conventional fight. Guerrilla operations are potentially of great importance in supporting tactical objectives.

3-153. Guerrilla forces can attack lines of communications, enemy supply depots and material, and isolated security forces to degrade the enemy's strength generally, and are particularly useful prior to and during invasions. Guerrilla forces can expect missions that directly assist combat operations of friendly tactical units. Since guerrilla forces occupy denied and enemy-controlled space with local area expertise, they can be particularly valuable in providing timely and detailed intelligence. Therefore, guerrilla forces can be employed as a reconnaissance and security force. They can mount diversionary attacks to support friendly operations elsewhere. Although guerrillas typically cannot hold and defend for long against strong opposition, they can temporarily seize key terrain in the enemy's rear area, such as key bridges, tunnels, defiles, dams, or other installations.

3-154. Although primarily of value in support of the tactical offense, guerrilla warfare can also assist friendly forces engaged in defensive operations, primarily through small-scale supporting offense actions. As major operations expand, evasion-and-escape operations expand to handle larger numbers of friendly personnel who may find themselves evaders. Psychologically, the impact of friendly conventional forces' success is magnified by intensified UW activity, and MISO should be crafted to exploit this. Finally, as conventional forces liberate previously enemy-controlled areas, linkup between friendly tactical commands and guerrilla forces usually takes place.

3-155. Tactical commanders who employ guerrilla forces must carefully consider their capabilities when assigning them operational tasks. Guerrilla units are usually organized as light paramilitary formations, and typically lack support structure, mobility, and communications. Typical guerrilla forces are composed of irregulars without formal military training and experience. Commanders should assume that guerrilla forces are not comparable to similarly sized conventional forces. Assignment of missions to guerrilla units should take advantage of their light-infantry characteristics and area knowledge. Army SF personnel working with the guerrillas are best suited to recommend to the tactical commander appropriate tasks for guerrilla forces.

3-156. The severest limitations common to guerrilla forces when employed with friendly tactical units is their shortage of adequate voice communications equipment and transportation. This is particularly true when guerrilla units are operating with a mobile force in a penetration, envelopment, or exploitation. For this reason, guerrilla units have a slower reaction time in terrain favoring a high degree of mechanical mobility. Conventional commanders may overcome this disadvantage by providing the necessary equipment or using the guerrilla force on an area basis. Another special consideration is the forward presence of guerrilla units behind the enemy's line of control. It is imperative that tactical commanders have SF liaison personnel on their staff to ensure establishment and adherence to no-fire areas where guerrilla forces operate.

Support of Ground Offensive Operations

3-157. As the JFC's area of influence overlaps the UW operational area, guerrilla units shift to operations planned to produce immediate effects on enemy combat forces. Initially, these activities are directed against enemy communication zones and army support troops and installations. As the distance between guerrilla and conventional forces decreases, guerrilla attacks have greater influence on the enemy combat capability. Guerrilla operations support penetrations and envelopments and are particularly effective during exploitation and pursuit.

3-158. Because of the high density of enemy combat troops in the immediate battle area, guerrillas can give little direct assistance to friendly forces in the initial phases of a penetration. Guerrilla forces can best support the attack by isolating, or assisting in the seizure of, the decisive objective. Guerrilla forces hinder or prevent movement of enemy reserves, interrupt supply of combat elements, and attack enemy command and communications facilities, fire support means, and airfields. Locations of critical installations and units that the guerrillas cannot effectively deal with are reported to the tactical commander for attack. As friendly forces near the decisive objective, guerrilla units direct their operations toward isolating the objective from enemy reserves. In some instances, guerrilla forces may be able to seize and hold the objective or key approaches to it for a limited time pending linkup with the conventional force.

3-159. Guerrilla units assist an enveloping force in much the same way as in a penetration. Guerrillas can conduct diversionary attacks to assist other forces' cover and deception plans. As in the penetration, guerrillas hinder movement of reserves, disrupt supply, attack command and communications installations, and reduce the effectiveness of enemy fire support. They may assist in containment of bypassed enemy units. They attempt to isolate the objective of the enveloping force. They may seize and hold critical terrain, such as bridges, defiles, and tunnels, to prevent enemy destruction. They may perform screening missions to the front and flanks or be a security element to fill gaps between dispersed units of the enveloping force. If used in a reconnaissance or security role, guerrilla units operate on an area basis; that is, they perform their security or screening role within a specified area during the time the enveloping force passes through the area. Guerrilla units usually do not possess the transportation or communications to accompany mobile forces.

3-160. As friendly tactical units pass from a successful penetration or envelopment to the exploitation of their gains, guerrilla operations increase in effectiveness. As the enemy attempts to reconstitute an organized defense or withdraw to new positions, he should be attacked at every opportunity by guerrilla forces. Enemy units previously assigned to rear area security duties are likely to be recommitted to attempts to restore a defensive position elsewhere. This may increase the vulnerability of rear area installations to guerrilla attack. Guerrilla forces can also assist in containing bypassed enemy units, can round up stragglers and prisoners, may be able to seize control of areas not occupied by the exploiting force, and contribute to the general enemy demoralization caused by the exploitation and subsequent pursuit. As linkup with the exploiting conventional force is accomplished, guerrilla forces and their underground and auxiliary counterparts can contribute to consolidation and transition activities.

3-161. Throughout coordinated use of guerrilla forces in support of major operations, operational control of the guerrilla force is retained at the level best able to coordinate the actions of the operation. SF working with guerrilla forces provide an excellent method of liaison and coordination with conventional forces. However, not all guerrilla forces will necessarily have SF or other coalition advisors. In such cases, as linkup becomes imminent, guerrilla units nearest the attacking force may be attached to or placed under the operational control of that force. Concurrent with linkup, responsibility for administrative support of the guerrilla force is passed to the tactical command.

Support of Airborne Operations

3-162. Guerrilla forces, by virtue of their location in enemy-controlled areas, can materially assist conventional or other forces engaged in airborne operations. They support airborne forces during the assault phase and subsequent operations. They may also be employed in conjunction with airborne raids and area interdiction operations.

3-163. Initially, guerrilla forces can provide selected current intelligence of the objective area upon which the airborne force commander bases his plans. Immediately prior to the assault, guerrilla units may be able to secure drop and landing zones, seize objectives within the airhead line, and occupy reconnaissance and security positions. Concurrent with landing of the assault echelon, guerrillas can conduct reconnaissance and security missions, provide guides and information, interdict approaches into the objective area, control areas between separate airheads and dispersed units, attack enemy reserve units and installations, and conduct diversionary attacks as a part of the cover and deception plan. Additionally, the guerrilla forces, auxiliary, and underground may be able to help control civilians within the objective area.

3-164. Correct timing of guerrilla operations with the airborne assault is essential. If committed prematurely, guerrilla forces may nullify the surprise effect of the operation and, in turn, be destroyed by the enemy. Conversely, if committed too late, the desired effects of the guerrilla force employment may never be realized. Army SF operating with UW forces are particularly effective in coordinating such guerrilla force activities with conventional operations.

3-165. As the assault phase of an airborne operation passes into the defensive or offensive phase, guerrilla forces continue to exert pressure on the enemy forces in the vicinity of the objective area. Guerrillas continue to provide up-to-date information on enemy moves and disposition. Attacks are directed against enemy units attempting to contain or destroy the airborne force, thus requiring the enemy to fight in more than one direction. Airborne forces that have an exploitation mission may employ recovered guerrilla units in reconnaissance and security roles as guides and to assist in control of void areas between dispersed units. If the airborne force is to be withdrawn, the guerrillas can assist to cover the withdrawal by diversionary operations conducted in the rear of enemy forces.

3-166. Guerrilla forces assist airborne raids in a similar fashion as they do the assault phase of an airborne operation. They provide information and guides, perform reconnaissance and security missions, and divert enemy forces during the withdrawal of the raiding force. An additional factor to consider before using guerrilla forces to support an airborne raid is the undesirable effect of enemy reaction on resistance organizations after withdrawal of the raiding force.

3-167. Airborne units are seldom committed to Normandy-like guerrilla warfare areas to conduct interdiction operations if the guerrilla force has the capability to conduct such operations. However, in areas where no effective resistance exists, or when supporting major invasions, airborne forces may be committed to conduct interdiction operations. Whatever guerrilla forces are located in areas selected for airborne interdiction, they assist the airborne force to conduct their operations. They provide intelligence information and guides, conduct reconnaissance and security missions, control the civilian population, assist in collecting supplies, and generally aid the airborne force commander in making the transition from conventional operations to guerrilla operations. SF detachments with experience in the operational area can prove particularly valuable in assisting the coordination of large airborne force guerrilla operations with those of the indigenous guerrilla force.

Support of Amphibious Operations

3-168. Guerrilla forces can also support conventional forces engaged in amphibious operations. Typical support activities include conducting operations to hinder or deny the enemy approach to the beachhead; seizing and holding all or a portion of the beach head; assisting airborne operations, which may be part of or complementing the amphibious assault; and conducting cover and deception operations to deceive the enemy as to the location of the actual beachhead. Guerrilla forces with adequate communications connectivity can provide current intelligence or spot for naval fires. Coastal defenses, obstacles, and radars would be typical targets in support of such amphibious assault.

3-169. Guerrilla forces operating within the objective area will probably be assigned to the operational control of the amphibious task force commander when he becomes responsible for the objective area. Such operational control will then probably pass to the landing force commander when he assumes responsibility for operation ashore. Depending on the situation, guerrilla forces may be advised and assisted by U.S. SOF. Such forces could be U.S. Army SF, but if they have been operating in the littoral areas of the UW area for some time, it is likely they have been developed and will be advised by U.S. naval or marine SOF conducting a UW mission. This is an example of the inherently joint nature of UW and highlights the imperative of effective joint planning and coordination. As in guerrilla force support to airborne operations, proper timing is critical to operational effectiveness and guerrilla force survival.

Linkup Operations

3-170. Most offensive operations in which guerrilla forces assist tactical commands involve a juncture between elements of the two forces. Normally during linkup operations, the guerrilla force is the stationary force, and the conventional unit the linkup force. However, not all guerrilla forces in an operational area are involved in linkup with tactical units. Some guerrilla units may be assigned missions assisting tactical commands where the requirements of the operation preclude physical juncture. For example, during a raid or area interdiction operations by airborne forces or when conducting operations as part of a cover and deception plan for an amphibious force, it is often undesirable to link up all guerrilla units with the attacking units.

3-171. Linkup, particularly between dissimilar combined combat organizations, can be dangerous. Therefore, careful coordination is required to mitigate risk of incident. Operational control of guerrilla forces supporting conventional operations should be made clear prior to linkup. The mechanics of linking up must not be confused by questions of who the guerrillas are to answer to. As in most other cases of operational synchronization, the attachment of Army SF to the guerrilla force prior to linking up provides an invaluable conduit for command, control, communications, and coordination.

3-172. The value of the trust SF build with their counterparts can be crucial in facilitating adherence to successful linkup procedures. Alternate but less certain methods of liaison preparatory to linkup include coordination through the underground, temporary exfiltration then reinfiltration of a guerrilla liaison team, or infiltration of conventional force liaison elements.

3-173. Linkup with guerrilla forces should not be assumed to be a one-time, one-place event. Guerrilla forces are typically dispersed throughout the operational area. Linkup may denote an entire series of relatively small events. Therefore, not only should coordination be made with the area command to have UW forces rally at predesignated points to facilitate linkups, but the conventional force commander must issue guidance for small-scale and unexpected linkup opportunity standard operating procedures.

3-174. Fire control lines and no-fire areas must be established to protect both the linkup force and the guerrilla forces from each other's fires. Guerrilla unit dispersion and the fact that civilian elements are a part of the UW force make these control measures particularly important. Communications coordination will be the key to successful linkup and minimizing of incidents. Here again, embedded SF units have the right communications equipment to provide liaison to the larger guerrilla forces. For smaller elements or individual combatants wishing to link up with conventional forces, turn-in procedures and near-recognition or identification measures should be disseminated through available joint force and resistance media channels.

GUERRILLA FORCE MISSIONS CONDUCTED AFTER LINKUP WITH FRIENDLY FORCES

3-175. In the event control of guerrilla forces is retained by the United States, missions may be assigned guerrilla forces after linkup with friendly forces has been accomplished. Depending on the political arrangements made with any resistance government-in-exile, operational control of guerrilla forces may be passed to tactical commanders. To avoid misuse of such light paramilitary forces, and to prepare for the challenges of amicable transition to postcombat activities, any attached SF detachments should remain with guerrilla units during this period. Conventional commanders must not overlook the new support requirements of these nonconventional forces.

Transition Considerations

3-176. Most guerrilla forces are irregular, paramilitary, and not suitable for standard conventional operations. However, properly coordinated guerrilla forces may be ideal for reconnaissance and security missions, screening the flanks of friendly forces, patrolling void areas between dispersed units, and providing guides. Some guerrilla forces may be used for light-static security duties in rear areas. Particularly capable and reliable guerrilla forces may be useful in counterguerrilla operations against enemy dissidents or newly coined resistors. The theater commander should be mindful of the role these guerrilla fighters may play in the challenges of transition to the newly formed indigenous government, and utilize, protect, or prepare them for that purpose.

Conventional Combat Operations

3-177. In some rare cases, properly trained and equipped guerrilla units can be employed as conventional combat units. Normally, they would require additional combat and logistical support, such as armor, artillery and transportation. A period of retraining and reequipping is usually necessary prior to commitment to combat. When so employed, they should be commanded by their own officers with assistance from SF advisors.

Reconnaissance and Security Missions

3-178. Because of their familiarity with the terrain and people in their operational areas, guerrilla forces possess a unique capability in a reconnaissance and security role. However, their lack of vehicular mobility and voice communications equipment are limitations on their employment with mobile forces.

3-179. The normal method of employment in reconnaissance and security missions is to assign guerrilla units an area of responsibility within which the guerrilla forces patrol difficult terrain and gaps between tactical units, establish roadblocks and observation posts, screen flanks, provide guides to conventional units, and seek out enemy agents and stragglers.

Rear Area Security

3-180. Guerrilla forces may be assigned rear area security missions. Such assignments should be based on their familiarity with and knowledge of the assigned area and populace. Typical areas to be secured include logistical and administrative installations, supply depots, airfields, pipelines, rail yards, ports, and tactical-unit trains areas. Guerrilla forces on rear area security may also patrol difficult terrain that contains bypassed enemy units or stragglers; police towns and cities; and guard lines of communications, such as railroads, highways, telecommunications systems, and canals. When provided with appropriate transportation, guerrilla units may be employed as a mobile security force reserve. Selected guerrilla, auxiliary, and underground elements may be effectively used in support of civil censorship operations conducted throughout the controlled area.

Chapter 3

Counterguerrilla Operations

3-181. Guerrilla forces are adapted by experience and training for use in counterguerrilla operations. Their knowledge of guerrilla techniques, the language, terrain, and population are important capabilities that can be exploited by conventional commanders engaged in counterguerrilla operations. Guerrilla forces may provide the principal sources of intelligence information about dissident elements opposing friendly forces. They have the capability of moving in difficult terrain and locating guerrilla bands. They detect guerrilla supporters in villages and towns and implement control measures in unfriendly areas. When properly organized and supported, guerrilla forces may be made completely responsible for counterguerrilla operations in selected areas.

Civil Affairs Assistance

3-182. Because of their specific area knowledge, cultural preparation, and linguistic capabilities, guerrilla forces or selected civilian support elements may be assigned to assist CA units. They may be directly attached to a special operations task force (SOTF)/JSOTF, combined joint special operations task force (CJSOTF), or conventional division-sized element. Guerrilla forces can perform the following:

- Assist in refugee collection and control duties.
- Assist in civil police duties.
- Help establish civil government.
- Recruit labor.
- Furnish or locate technicians to operate public utilities.
- Guard key installations and public buildings.
- Assist in the review and censorship of material for dissemination through public media facilities.
- Assist in restoring the area to some semblance of normality.

3-183. Shrewd commanders will anticipate the utility of transitioning guerrilla forces from a combat experience into adjuncts of new government authority, and will plan for such forces' reintegration, organization, compensation, and accountability.

INTELLIGENCE OPERATIONS

3-184. Insurgencies and resistance movements rely on accurate and timely intelligence for survival and effectiveness like any other political actor with the will to use violence. The area command is not, however, primarily an intelligence agency but a military force. The intelligence system of the operational area is primarily geared to support the command itself. The primary intelligence interests of the area command will be obtaining and acting on intelligence that supports its own insurgent tactical operations and political activities. However, when an insurgency or resistance movement gains U.S. support as part of a UW campaign, such intelligence as can be collected by the indigenous resistance forces can be of tremendous value to the combatant commander. Definitions for intelligence operations terms are as follows:

- *Intelligence operations*: The variety of intelligence and counterintelligence tasks that are carried out by various intelligence organizations and activities within the intelligence process. Intelligence operations include planning and direction, collection, processing and exploitation, analysis and production, dissemination and integration, and evaluation and feedback. (JP 2-01, *Joint and National Intelligence Support to Military Operations*)
- *Intelligence preparation of the battlespace*: The analytical methodologies employed by the Services or joint force component commands to reduce uncertainties concerning the enemy, environment, time, and terrain. Intelligence preparation of the battlespace supports the individual operations of the joint force component commands. (JP 2-01.3, *Joint Intelligence Preparation of the Operational Environment*)

Activities

Note: Like PE, UW intelligence operations are highly sensitive and current examples will almost always be classified. Therefore, the examples given in this section rely on the classic era of UW—WWII—to illustrate some of the constituent ideas. Although dated, these examples provide insight into the types of intelligence operations which have been and could again be conducted. Presentation of intelligence operations techniques in this publication is not a statement that U.S. forces necessarily conduct or support any one of them, nor should such U.S. conduct or support be inferred.

TACTICAL INTELLIGENCE

3-185. Tactical intelligence which allows the insurgent or resistance organization to survive and to take action against the opponent is the priority. Insurgent intelligence assists associated guerrilla forces by providing valuable data about the enemy and the area of impending combat. This may include the number of enemy troops, their deployment, their unit designations, the nature of their arms and equipment, the location of their supply depots, the placement of their minefields, the pattern and routine of their patrols, the morale of the troops, and various topographical factors, such as swamps and ravines, that govern access to enemy emplacements. This information will be used by the insurgent or resistance organization itself for its own tactical planning. Every member of a resistance is an information provider.

3-186. The location of the area command in enemy-controlled territory also makes available to the theater commander an additional means of developing intelligence generally unavailable to other theater forces. Some examples include—

- Finely detailed and timely order of battle intelligence data.
- Indigenous information crucial to the development of effective insurgent psychological warfare activities.
- Information of political, sociological, and economic intelligence value at a resolution and vantage otherwise unavailable.
- Precise, accurate, and timely target information for tactical and strategic air forces plus post-strike information.
- Intelligence data to support specific tactical operations, such as airborne, amphibious, or armored operations.

3-187. Sometimes this information is obtained directly by underground personnel by visual observation of the targets. For example, members of the French resistance reconnoitered German coastal defenses in the preparation for the Allied invasion of France in June 1944. Such data may also be collected by the local populace, or "popular antennae," as these sources are described in one Vietminh manual. The Vietminh used children playing near French fortifications as a source of information on troop arrivals and departures, the guard system, and other pertinent details, which aided the guerrillas in planning attacks—all of which were easily observable by untrained children.

3-188. In UW, intelligence training and guidance by the sponsoring power can improve the quality of such information because most undergrounds initially lack the personal experience in this type of work. For example, the WWII Jedburgh teams in France trained resistance workers in their intelligence surveys and helped to coordinate their collection efforts. Likewise, Red Army personnel were assigned to the Soviet partisans to direct their activities. In the modern context, U.S. Army SF and other governmental agencies can provide this kind of training and guidance.

3-189. A Soviet handbook distributed to partisans for use in regions under German occupation instructed that... "If you happen to encounter troops . . . do not show that you observe the enemy . . . ascertain the color of their headgear, their collar braid, and the figures on their shoulder straps. If they have questioned the inhabitants about something, try to find out what the Fascists have asked." The guidebook also gave tips about ascertaining enemy intentions: "if an attack is planned, trucks will arrive loaded and depart empty; if the enemy intends to retreat fuel and foodstuffs will be removed, roads and bridges will be demolished, telephone wires will be removed, and trains and trucks will arrive empty and depart full."

INTELLIGENCE IN SUPPORT OF SABOTAGE

3-190. Reconnoitering transportation and communication facilities prior to sabotage attacks occupied much of the time of French resistance persons. Often working closely with Allied advisors, these people surveyed targets earmarked for sabotage on D-day. In reconnoitering a bridge, for example, resistance members looked for such factors as the guard system covering the bridge—if a number of permanent troops were evident, a step to eliminate them had to be included in the sabotage plan; when there was only an occasional patrol, the resistance would time an attack to avoid the patrol. Observers also noted and reported the characteristics of bridge construction so that the size of the explosives could be calculated.

3-191. By determining the schedule of enemy train movements, saboteurs were able to destroy stretches of railroad track while it was in use, thereby compounding the wreckage and complicating repair work. Danish railroad saboteurs had an elaborate system to provide this information. Throughout Jutland, underground members were stationed near major terminals to note the departures of enemy troop trains. Whenever one was seen, the observer telephoned prearranged code phrases to the sabotage cell in the town next on the railroad line. Members of this cell then proceeded to predetermined spots on the tracks to lay their mines. With this advance notice, the mines could be placed at the last moment, preventing detection by patrolling guards. The train delayed by sabotage might eventually reach the next stop, but observers there would be waiting to repeat the process. Using these observers along a train's route, the resistance was sometimes able to slow a train's progress by days or even weeks.

3-192. Production facilities were also surveyed by undergrounds in preparation for sabotage attacks. When possible, underground personnel are often aided in planning factory sabotage by outside intelligence experts, for these are best qualified to make the necessary technical judgments. It is a problem in itself to determine just which components in a plant should be incapacitated. Prior to the blowing up of a Norwegian heavy-water plant being operated by the Germans during WWII, the preliminary reconnaissance was done by an SOE agent who parachuted into Norway. Details about the factory's equipment were obtained from a Norwegian scientist in London. Other data, perhaps about the guard system and access to the equipment, apparently were supplied by underground workers in the plant.

INTELLIGENCE FOCUSED ON SCIENTIFIC AND MILITARY SECRETS

3-193. Secret scientific and military data can be obtained by recruitment of employees in scientific and military installations, or by simple observation elsewhere. An example of the latter was the valuable data about the V-2 rockets obtained by the Danish resistance. During the summer of 1943, fishermen near the island of Bornholm began to report the rash of unidentified objects in the sea; they were recorded by the resistance leader on the island. In August, the island's police commissioner notified the underground leader of the crash of a flying craft in a nearby field. The two men rushed to the scene before the Germans and found the wreckage of what was clearly a new kind of aircraft. The only identification mark was a number: "V1-83." The men took photographs immediately before the arrival of the German investigators. From the skid marks the underground leader was able to determine that the device had come from the southwest. From the pictures, the underground chief drew a complete sketch of the weapon. This sketch, the photographs, and the notations as to the direction from which the missile had come were sent by courier to England, providing the British with perhaps their first technical data on the new German rocket.

POLITICAL INTELLIGENCE

3-194. Underground agents also collect political intelligence. They note the statements and activities of persons to determine who favors the regime, so that these persons may be closely watched or eliminated if their actions seriously threaten the underground. In Belgium during WWII, the Movement National Beige kept files on collaborators and campaigned by threatening phone calls and letters to dissuade these individuals from working with the enemy. If this failed, the collaborators were often executed. The list of collaborators was never made public in order to keep concealed the extent of cooperation with the enemy.

3-195. In wartime, the underground also notes the morale of the enemy soldiers. The Polish Home Army systematically collected data on German troops by reading their mail. There were too few Germans to handle all of the postal work; thus, many Poles were employed. These workers would open letters and photograph the contents before sending them on. From these letters, a fairly good estimate could be made of the enemy's morale.

The Value of Insurgent/Resistor Intelligence in Unconventional Warfare

"There is no doubt that most partisan actions inflicted damage upon the opposing forces. Some of the damage was severe.... [However,] their second great contribution was in the field of intelligence.... [It] cannot be doubted that the partisans served well as field intelligence, especially after Army intelligence officers had been seconded to all partisan staffs in 1943. The scope was wide—the partisans were everywhere—their location ideal—behind the enemy's front—and their instructions were detailed—in the Field Service Regulations, the Partisan Handbook, the Guide Book for Partisans, and so on.

We can be almost certain that again and again Russian attacks were mounted in those areas which partisan reports had indicated as vulnerable. The Russians during the war became expert in attacking the enemy's weakest points: the small front-line gaps in the winter of 1941-2, the front held by German satellite troops at the beginning of the Stalingrad battle; and if there was neither gap nor satellite, it was almost always the seam between two enemy formations which the Red Army selected for its breakthrough attempts.... There was only one source which could consistently direct the Red Army against the weakest link of the enemy front, and this task... was entrusted to the partisans.

We are of course better informed about the value of French partisan intelligence. 'In fact, the day the battle (in France) began,' says General De Gaulle, 'all the German troop emplacements, bases, depots, landing fields and command posts were precisely known, the striking force and equipment counted, the defense works photographed, the minefields spotted.... Thanks to all the information furnished by the French resistance, the Allies were in a position to see into the enemy's hand and strike with telling effect.'

These words speak for themselves; no finer testimonial could be given."

Otto Heilbrunn, *Partisan Warfare* (1962)

ADDITIONAL CONSIDERATIONS

3-196. The underground organization and many of its activities are based upon a fail-safe principle so that it is organized and if one element fails, the consequences on the total organization will be minimal. This is especially true for intelligence cells. Almost all clandestine organizations that are susceptible to compromise by security forces have parallel organizational units and networks of units. In every case, the underground attempts to have a backup unit that can perform the same duties as the primary unit if the latter is compromised. It usually takes a long time to establish a unit or net and the underground must plan for contingencies, such as the compromise of the primary unit or increased government security measures. Thus, the organizational expansion of undergrounds is usually in a lateral direction by duplicating units and functions. The decentralization extends to all functions.

3-197. If the cell operates as an intelligence unit, its members may never come in contact with each other. The agent usually gathers information and transmits it to the cell leader through a courier or mail drop. The intelligence cell leader may have several agents, but the agents never contact each other and only contact the cell leader through intermediaries. The underground usually does not jeopardize intelligence units by demanding that they perform sabotage as well, for sabotage operations may draw attention to individuals and compromise their usefulness as intelligence agents.

3-198. Some intelligence tasks call for the recruitment of persons with access to important information. Gathering intelligence is a major activity of an underground, and to do it effectively an underground contacts and develops sources in all areas of interest. The range of persons so utilized may be wide, varying from a peasant woman living near an enemy convoy route to a person in a sensitive governmental bureau. Immediate access to valuable information is not the only criterion for selection, however. Sometimes the underground recruits individuals to provide information which is valueless in order to keep them on a string, in the hope that they may advance in their profession and be in a position to supply useful information someday. If time permits, it may be better to wait for someone to work his way up into a position of trust rather than to seek to recruit someone in a high position with the attendant risk of failure or compromise.

3-199. The unusual factors of large-scale, illegal immigration and the special political status of the mandate in Palestine created a unique situation for an underground to operate in. Drawing on former members of WWII undergrounds, the Haganah developed an effective intelligence network. It also transcended national boundaries and used the sympathy of other nations for the refugees to aid the movement.

Chapter 4
Supporting Activities and Legal Considerations

Build for your team a feeling of oneness, of dependence on one another and of strength to be derived by unity.

Vince Lombardi

Although UW is most closely associated with Army SF, SF do not conduct UW unilaterally and without support from other agencies and elements. During the conduct of preconflict shaping and engagement activities particularly, UW must be conducted in the context of normal U.S. statecraft and in conjunction with USG interagency partners. Should UW be employed in support of major operations, careful coordination with, and mutual support of, conventional forces will be required. In all cases of UW, integration of the affiliated capabilities of MISO, CA, and the full range of logistics support will be essential. Moreover, as UW often inhabits the gray area of political engagement blending into warfare, careful attention to legal considerations and respect for authorities is crucial.

INTERAGENCY SUPPORT

4-1. USG interagency support for a UW campaign is essential from the earliest national or theater command interest in exploring the viability of such an effort. In almost all instances, the target regime of a proposed UW campaign will be in a country not in a declared state of war with the USG and its allies. Cases of straightforward reaction to obvious aggression such as the Iraqi invasion of Kuwait in 1990 or the classic WWII response to Axis occupation are the exceptions. Most often, the U.S. Ambassador responsible for the target country—whether there is official U.S. representation resident in that country or not—will be involved in the earliest assessment and decisionmaking over activities to be conducted in his country. In fact, most of the routine, earliest, and most-highly sensitive activities that will be conducted in a target country will be done by agencies of the USG other than the DOD. It is probable that a tasking for a theater commander to begin initial assessment of UW potential in a country will be based on sensitive evaluations that have already been conducted in the AO by operatives other than the military.

4-2. Whatever the origin of initial UW planning, it is imperative that planners and commanders develop a collaborative framework with relevant agencies outside of the DOD when formulating options. Often, such collaborations will seem ad hoc based on the conditions unique to the circumstances. The theater commander may call for the creation of an interagency task force which will almost certainly contain representatives from the Department of State (DOS) and the intelligence community at a minimum. Factors such as political sensitivity, scope of the challenge, expected duration, signature limitations, threat vulnerabilities to a range of indirect applications of U.S. national power instruments, and so on are examples of variables which will combine to suggest the appropriate mix of interagency participation.

4-3. Finally, although UW is "warfare" and therefore an option available to national decisionmakers through the military instrument of power, the design of UW campaigns methods is as much "art" as "science." The imaginative commander will want to take advantage of the many indirect ways of applying coercive and disruptive power through departments and agencies not under his direct control. The commander and staff must collaborate with the interagency. Moreover, although JP 3-05 and other sources make clear that military forces committed into a foreign AO remain under the control of the GCC, the ambassador remains the president's personal representative to that area; especially in the earliest phases, failure to cooperate constructively with the country team would be counterproductive.

Chapter 4

CONVENTIONAL FORCES SUPPORT

4-4. In virtually all cases, SOF commanders will need to obtain support from and conduct coordination with conventional forces. SOF rely on the Services for a variety of logistics and sustainment support. Conventional forces may, in some instances, be appropriate to provide quick-reaction forces, medical evacuation, transportation, logistics, reconnaissance and surveillance, or fire support for SOF during UW operations. Such support will be usable most often when attribution of support is not an issue.

4-5. Conventional forces may be able to support a UW campaign indirectly by securing sanctuary areas in a neighboring country that permit SOF or other agencies to train resistance personnel who have been temporarily exfiltrated from the target country. Conventional forces may also support indirectly by serving in a role that diverts attention and resources from the target country or its allies.

MILITARY INFORMATION SUPPORT OPERATIONS

4-6. MISO—and much more broadly, the full range of U.S. information and influence activities to shape perceptions—will always play a central role in successful UW. The earliest practical synchronization of efforts between the GCC and the chiefs-of-mission in the area is essential. MISO planning must be involved at the earliest stages of planning and should prepare the indigenous population prior to the insertion of SOF. The information plan should include desired effects, proposed actions, end state objectives, supporting objectives, and the requirements for each phase of the campaign. The MIS plan should also outline key messages and themes for different audiences during each phase. MISO will be fundamental to every phase of a UW campaign.

4-7. One of the most salient characteristics of UW conducted in the 21st century compared to previous eras is the widespread use of electronic devices worldwide. This development creates additional opportunities to shape the operational environment through both direct influence operations to individuals and exploiting vulnerabilities in regime information architectures. UW planners must consider exploiting the full range of recent developments in the cyber environment. Cyber infrastructure questions will include—

- What portion of the population has Internet access?
- What other telecommunication media exist?
- How effective are the target regime's censors and controls, and what are their weaknesses and vulnerabilities?

4-8. Finally, the state of development of such useful technology will often outpace the doctrine and procedures intended to exploit it. Planners must integrate cyber warfare subject-matter experts into unconventional campaign planning at the earliest practical opportunity, both to assist in developing exploitable cyber and electronic warfare options and to identify authority and approval gaps at the earliest possible instance. U.S. Strategic Command, U.S. Cyber Command and other DOD organizations or OGAs will be typical partners in modern UW planning.

CIVIL AFFAIRS OPERATIONS

4-9. CA forces are particularly suited to execute UW. As opposed to the broader purposes of CMO as conducted by conventional forces, CAO conducted by CA are uniquely task-organized for the execution of all seven phases of a UW campaign. Through focused planning, assessment, and targeting, CAO identify and exploit the civil center of gravity in favor of a guerrilla force and the supported shadow government, with the ultimate goal of collaborating with interagency and guerrilla/legitimate government partners to ensure stability and a stable civil infrastructure. While CMO are also conducted by SOF and conventional forces, its purpose is to address more general issues affecting civil infrastructure, allocation of resources, support of civil institutions, and overall mitigation of the impact on the civil domain made inevitable by the presence of military activity. While CA forces are adept at aiding transition from a wartime guerrilla campaign against a regime to a stable post-conflict reconstitution and reconstruction of the UW operational area, it should be noted that due to restraints in manpower, time, and material, CA forces cannot take the lead on transition, but are value added to relevant government/nongovernmental organizations, and the host-nation government.

4-10. CA operational support during each phase of UW may include the following:
- Phase I:
  - Providing detailed civil considerations analysis of the JSOA to identify potential civil vulnerabilities, grievances, and weaknesses within civil society that may be leveraged by the insurgency to generate mobilization of the population.
  - Identifying foreign humanitarian assistance funding sources.
  - Initiating CA support coordination with any government-in-exile or shadow government representatives if established.
  - Initiating transition planning by developing a disengagement concept and identifying the CMO end state.
- Phase II:
  - Integrating with the pilot team-planning cell.
  - Identifying sources of foreign humanitarian assistance, to include intergovernmental organizations and nongovernmental organizations.
  - Preparing to insert with the pilot team, if required, to support initial assessments.
  - Preparing to provide detailed CAO or CMO analysis of pilot team assessments.
  - Identifying quick, high-impact projects based on pilot team assessments.
- Phase III:
  - Validating CA inputs to the intelligence preparation of the environment.
  - Identifying and liaising with key leaders within the indigenous population who may influence the CAO/CMO plan.
  - Preparing to insert with the SF detachments if required.
  - Conducting key leader engagement.
- Phase IV:
  - Refining CA input to intelligence preparation of the environment.
  - Facilitating the buildup of the resistance force.
  - Initiating CAO (foreign humanitarian assistance, nation assistance, and populace and resources control).
- Phase V:
  - Validating measures of effectiveness.
  - Monitoring or assessing effects.
  - Executing CAO.
- Phase VI:
  - Mitigating impact of combat operations on the indigenous civilian population.
  - Surging foreign humanitarian assistance.
  - Preparing to execute planned populace and resources control (dislocated civilian operations).
  - Deconflicting intergovernmental and nongovernmental organization operations supporting unity of effort.
- Phase VII:
  - Executing support to civil administration operations to advise and assist the new or provisional government.
  - Supporting posthostility institutions to foster legitimacy and transparency of government.
  - Supporting the interagency execution of strategic and operational stability operations.

LOGISTIC CONSIDERATIONS

4-11. UW logistic support is different from support to other types of operations and even other types of special operations. This difference derives from the requirement to assist indigenous forces in denied areas typically characterized by a politically sensitive and physically challenging environment and usually over long time periods. Theater sustainment architecture will typically be undeveloped, especially when UW is used as a strategic initiative or response without planned U.S. major operations.

4-12. UW campaign planners need to consider key characteristic challenges of sustaining UW, have an overview of the most typical mode of resupply, understand select particulars of how supplies are received in the JSOA, and be familiar with the responsible organizations and echelons that will be involved in UW campaign planning for logistics.

UNCONVENTIONAL WARFARE CAMPAIGN SUSTAINMENT PLANNING

4-13. UW will involve activities and missions that require logistics support conducted with low visibility and under clandestine conditions, and beyond the abilities or authorities of standard logistics. This type of support is referred to as unconventional logistics. Generally, logisticians providing such support remain cognizant of conventional logistics principles, such as supply chain management, but must adapt existing— or develop new—tactics, techniques, and procedures to deal with unusual processes and requirements for acquisition, storage, funding, and transportation. Logisticians involved in supporting UW use conventional providers within the DOD as available and appropriate, but are not restricted to those providers. Demand for such activity can result from missions representing the highest political sensitivity and which are conducted across the entire spectrum of military operations.

4-14. Sustainment planning for any operation involves the identification of requirements, the provision of organic capabilities to meet those requirements, the subsequent identification of shortfalls, and consideration of leveraged or acquired options to mitigate those shortfalls. This is the conventional logistics estimate model and is applicable for the UW concept of support planning as well. However, because of operations security requirements and the constraints and limitations inherent to UW, challenges exist in the unconventional logistics arena that requires increased diligence in the planning processes. Some of the key challenges include the following:

- The unique requirements, difficulties, and nonstandard authorities and funding mechanisms of outfitting indigenous irregular forces and population groups in a manner that does not raise the signature inside or outside of the AO, and in a manner which is often both covert and clandestine over long time periods, potentially many years.
- Joint force logisticians' challenge to coordinate joint force supply and sustainment with the ongoing operations of non-DOD departments and other USG agencies while maintaining the highest level of operational security.
- Joint force logisticians' challenge to manage and integrate the supply and sustainment of all Service components' needs over long time periods.
- Establishing an intermediate staging base is a critical step in successful unconventional logistics. The intermediate staging base should have experienced unconventional logistics logisticians, resource management personnel, contracting specialists for rapid acquisition of goods, and direct access to nonstandard aircraft (or other platforms) to enable timely resupply.
- The challenges unique to maintaining supply lines to deeply isolated elements through denied territory and over long time periods. Otherwise simple procedures such as the repair of sophisticated radios or emergency surgeries or even mortuary support become magnified logistical challenges at the end of a precarious supply line.
- There are a very limited number of professional logisticians properly trained to understand the unique requirements of a UW campaign; this is potentially a significant campaign planning limitation. Experienced unconventional logisticians are required to properly anticipate logistic branches and sequels to campaign plans.

Supporting Activities and Legal Considerations

- Due to the imperative to remain covert and/or clandestine and the difficulties inherent in denied areas and difficult terrain, it is usually easier to infiltrate most personnel and materiel discretely than it is to exfiltrate those same personnel and materiel discretely.
- The need to stockpile potentially significant quantities of both standard and nonstandard items for anticipated scheduled, on-call, and emergency resupply in a manner that is both clandestine and financially sustainable, especially when responding to changing conditions and updated indigenous information. Because it is difficult to anticipate every future circumstance in highly unorthodox situations, it is extremely important for planners to secure contingency contracting authorities and establish contingency procurement procedures in conjunction with mission execution orders.
- Finances do not always merely support other operations. Sometimes, application of finance to select indigenous actors/institutions is itself the operation. Such "money as a weapons system" activities need to be anticipated as far in advance as possible, and the supply of appropriate stocks of money becomes a sustainment challenge.
- Signature management is a serious challenge. The more sensitive the operation and the larger the commitment of U.S. assets to the operation, the greater this challenge will be. Planners must consider the second- and third-order effects of primary actions and inputs. For example, UW campaign plans that rely on local procurement in or near the AO must allow for perceptible changes to local economic patterns. Sudden increases in local currencies will present changes detectable to threat authorities. Changes to ground transportation, flight, shipping, or storage patterns or volumes; hiring anomalies; indigenous diet changes; and noise and light increases in unusual places are all examples of the unconventional logistics signature management challenge.

4-15. The key elements of a good concept of support involve detailed planning with respect to providing goods (supplies and equipment), services, and support (construction and facilities). Conventional operations have established funding streams, supply chains, and players operating support nodes.

4-16. By contrast, unconventional logistics support often lacks the up-front structure. Once the unconventional logistics requirements have been identified, the unconventional logistics planner must consider how the required goods, services, or support needs will be funded. The unconventional logistics planner will then identify the supply chain or procurement trail and understand that they may be different for each item within the same operation. The identification of players in the operation often includes a METT-TC determination of the point at which a professional sustainer will be involved rather than using an SF element. It could also involve determining what conventional or unconventional systems will be leveraged to provide the required support.

4-17. The unconventional logistics support planner must anticipate, monitor, and respond to the requirements associated with these changes. Time is more of a critical planning consideration for the unconventional logistics sustainment planner than it is for his conventional counterpart. It sometimes takes longer than anticipated for the SF element to receive the required goods and services based on secure operating requirements, length and complexity of nonstandard supply chains, and the process of obtaining authorities for funding. Unconventional logistics places extraordinary demands on both the human resource system and the Soldiers assigned to conduct these types of operations. Supporting UW requires maximum flexibility, financial and contracting agility, and rapid operational and strategic reachback to provide responsive support.

SUPPLYING THE AREA OF OPERATIONS

4-18. Logistic support to UW is broader than just supply, especially from the perspective of the TSOC or the JSOTF. However, UW campaign planners must appreciate the vital lifeline that supply represents to the executing units in the AO. Although Army SF and other special operations units are adept at improvisation and living off of the land, practically speaking there is still no substitute for reliable and timely supply of such items as ammunition, advanced medicines, and sophisticated communications equipment to the UW battlefield. TC 18-01, *Special Forces Unconventional Warfare*, discusses supply in more detail.

Chapter 4

4-19. There are two main categories of supply: accompanying supply and external resupply. External resupply is comprised of automatic, emergency, and on-call (or routine) resupply. The following sections discuss each of these categories and subcategories.

Accompanying Supply

4-20. The SFODA may take accompanying supplies into the JSOA at the time of infiltration. The detachment receives these supplies in isolation at the JSOTF or special operations task force (SOTF). While undergoing mission preparation in isolation, the detachment prepares and rigs accompanying supplies for delivery in conjunction with infiltration. This preparation must include packaging and load consideration to facilitate transportation subsequent to infiltration. The situation may dictate that these supplies are cached following infiltration for later use. The threat in the JSOA dictates the quantity and type of supplies and equipment the detachment can include. Other influences are the—

- Capabilities, size, and responsiveness of the guerrilla force to sponsor assistance.
- Enemy capabilities and situation.
- Method of infiltration (air, land, or sea).
- Requirements for survival, evasion, resistance, and escape.
- Available resources in the JSOA.
- Size and capability of the reception committee.
- Requirements for sustaining operations pending receipt of an automatic resupply.
- Need for key items of equipment to partially equip a cadre nucleus of the guerrilla force when the SFODA expects a reception committee upon infiltration.
- Other items of equipment and supplies to help establish rapport with the guerrillas.

4-21. To facilitate isolation and field planning, the TSOC and subsequent JSOTF must make detachments aware of available materiel. These elements are responsible for establishing a catalog of available materiel for a campaign effort. This catalog will include standard and nonstandard materiel. Detachments need to understand if quantities have already been allocated and the given quota for the number of resupplies they can expect. They also need to know whether resupply scheduling is by month or by phase (air, sea, and ground). The willingness to assume risk with platforms to resupply forces may increase as phases progress toward initiation of combat operations.

External Resupply

4-22. External resupplies are procured and delivered to the JSOA by the sponsor (JSOTF), based on the needs of the resistance force or insurgents, as well as the detachment. Resupply is planned in isolation to be delivered after infiltration at a coordinated location and time automatically (automatic), as requested (on-call or routine), or based upon a no-communications trigger (emergency). The detachment preselects resupply items and delivery merchandise during isolation to replenish or supplement supplies its members consume or to fulfill other requirements. The detachment receives these items after infiltration.

Automatic Resupply

4-23. The detachment plans for automatic resupply before infiltration, coordinating for the delivery time, location, contents, and the identification marking system or authentication. Automatic resupply is delivered after the detachment successfully infiltrates and establishes radio contact, unless the detachment cancels, modifies, or reschedules the delivery. Automatic resupply augments supplies or equipment that the detachment could not carry in on the initial infiltration or were not required until a later date. Automatic resupply also reinforces U.S. support of the guerrillas. In addition, it fulfills the need for selected items to equip a nucleus of the guerrilla force should accompanying supplies be unrecoverable during infiltration.

4-24. The JSOTF or SOTF schedules the delivery of automatic and emergency resupply to the detachments. Preplanned automatic resupply provides the guerrilla force with immediate supplies and equipment until on-call or routine resupply procedures are established. Supply personnel normally pack equipment and supplies in appropriate aerial delivery containers that have a cargo capacity of 500 pounds or less to ease handling and transportation within the JSOA. Packers mark door bundles for easy identification once they arrive on the drop zone. To allow rapid clearance of the drop zone, personnel ensure the contents of each container are in man-portable units of about 50 pounds each. Packers must brief drop zone parties on these man-portable containers. If personnel must carry the containers long distances, the detachment must arrange transportation assets with the guerrilla support arm (auxiliary). The detachment can cancel, modify, or reschedule automatic resupplies, depending on their requirements.

Emergency Resupply

4-25. The purpose of the emergency resupply is to provide essential equipment and supplies in order to restore operational capability and survivability of the detachment. Typical items contained in the bundle may be communications equipment, batteries, weapons, ammunition, money, and handheld global positioning systems. A coded message, a radio request, or the absence of any detachment communication over a prearranged period can trigger an emergency resupply. The detachment and the supporting HQ must clearly understand the sequence of events, time required, and assets available to deliver the emergency resupply.

4-26. Although detachments can request special items, it is highly likely that all emergency resupply bundles will contain generic items in order to support numerous detachments. As a minimum, resupply should consist of communications equipment and enough mission-essential supplies to establish base contact.

On-Call or Routine Resupply

4-27. When the detachment establishes communications with the JSOTF or SOTF, external supply begins on call. Personnel use the abbreviated code of a catalog supply system contained in the signal operating instructions to request supplies based on operational need. These supplies consist of major equipment items that units do not consume at a predictable rate. Theater Army area command depots, the JSOTF, or the SOTF hold these items in readiness for immediate delivery on a specific mission-request basis.

4-28. To determine the quantity of supplies to request, the detachment considers the rate of expansion of the guerrilla force, the anticipated tempo of operations, and the detachment's ability to receive, transport, store, and secure incoming supplies.

4-29. The detachment also anticipates its operational needs for supplies and equipment in the JSOA. The mission operation cell at the JSOTF or SOTF packs and rigs the supplies into man-portable loads and color-codes them before infiltration. The mission operations cell color-codes the supplies IAW the type of supplies in the load so personnel need not open them for identification.

4-30. As the guerrilla force expands and logistic requirements increase, internal popular support will lose its ability to provide subsistence for the resistance without creating hardships for or lowering the living standards of the civilians. At this point, the force must obtain logistics support from an external source. This dependence on the external source requires a routine supply system. As the JSOA grows, the need for external supply normally outgrows the on-call method of requesting supplies.

ORGANIZATION OF SUPPLY IN THE AREA OF OPERATIONS

4-31. The area command, with advice from U.S. advisors, plans, develops, operates, and controls the guerrilla force logistics system. Each element of the area command has a specific role in the logistics system, and the force develops each system to meet the specific needs and peculiarities of the JSOA. Elements may centralize or decentralize supply organization systems. During the organization and buildup phase, the command may centralize logistic operations, moving all supplies into one collection area. The detachment procures logistics items from throughout the JSOA, processes them through a centralized or decentralized collection point, and distributes them to all units of the guerrilla force.

4-32. As the JSOA matures, subordinate units take over a sector in which they are responsible for establishing a separate and decentralized supply procurement system. They distribute all supplies throughout the JSOA to several distribution centers. This decentralization improves security because compromise or destruction of the procurement system in one sector will not destroy the entire apparatus. Another advantage of this system is that it permits an equitable distribution of the logistics burden on the civilian population. Elements keep movement of supplies between sectors to a minimum, and personnel do not pass names, storage sites, and caches from sector to sector. The area commander delegates supply operations to sector commanders. He retains the responsibility for the overall plans. He also reports supply needs to the sponsoring power and issues directives covering operations. Plans and directives may include—

- Organization of supply and service support units.
- Organization and employment of civilian support units.
- Systems of levy and barter on civilians.
- Receipt of payment for supplies.
- Collection, storage, transportation, and distribution of supplies.
- Quantity and type of supplies personnel are to maintain.
- Allocation of supplies to major lower commands.

4-33. The area commander provides all supply items to the sector commander. The sector commanders supply their units and conduct supply operations according to the plans, directives, and orders of higher HQ. Individual units within their assigned sectors conduct decentralized supply operations. The sector commander makes his needs known to the next-higher HQ for supplies and equipment not available within his area. He distributes all supplies and equipment received from higher HQ. Besides supplying his sector, higher HQ may charge him with supplying adjacent sectors.

Delivery Systems

4-34. The SF detachments in the JSOA deliver all external supplies and equipment to the guerrilla force. It is essential that personnel control access to sensitive items, such as weapons, ammunition, demolitions, radios, drugs, or special equipment. A detachment member must be present at all deliveries of external supplies to ensure positive control and accountability.

Delivery Means

4-35. The preferred mission delivery method for external resupply is by sponsor aircraft, surface ship, or submarine. At first, planners may determine aerial delivery by parachute is the best means of supply to a JSOA. Personnel may use free-drop techniques for certain hardy items. Later, as the JSOA expands and comes under greater friendly control, detachment members use air-landed supply missions. Supply personnel normally use surface ships or submarines when JSOAs are next to waterways or seas. Resupply operations require secrecy to protect the resupply platform and the reception element. Personnel normally conduct these operations during limited visibility.

Packaging

4-36. The JSOTF or SOTF support center personnel prepare supplies and equipment for delivery to a JSOA. The size of the package and the number of packages determines the delivery means. Personnel base the packaging system on man-portable packages weighing about 50 pounds. With this weight limitation, members of the reception committee transportation party can easily move the packages from the supply point to safe sites. Personnel equip man-portable packages with carrying straps and mount the packages on pack boards. The transportation party color-codes the packages so personnel can easily identify their contents upon arrival. The JSOTF or SOTF support center personnel ensure each package is—

- Waterproofed to permit aboveground and limited underground or underwater cache.
- Packed with instructions (in the indigenous language) for all equipment.
- Marked with a prearranged code to identify the contents.
- Packed with an inventory list to aid in identifying lost or damaged materiel.
- Packed to protect sensitive communications and medical items by using clothing, blankets, or other padding.

4-37. The JSOTF or SOTF support center personnel may put combat and morale supply items in the same container. They package ammunition and cleaning equipment with weapons and batteries with flashlights. Personnel may include additional small arms ammunition, by caliber, as an individual package.

Supply Procedures

4-38. Detachment members use a catalog supply system code to accelerate on-call resupply requests and ensure accurate equipment identification and supply items. The system also reduces radio transmission time. To permit maximum user flexibility, the system identifies single major equipment items or several associated items by code words. Personnel catalog these items by class of supplies and group them in individual packaged items or several associated unit items packed together. The catalog supply system is not secure, but it reduces message length and transmission time when personnel request a variety of supplies. The catalog supply system is—
- Based on mission requirements, concept plans, and standard operating procedures.
- Prepared under the supervision of the group logistics staff section.
- Reproduced in miniature form for operational missions and published in the signals operating instructions by the group signal officer.

LOGISTICS RESPONSIBILITIES

4-39. Due to the typically sensitive political nature, indigenous focus, and long-term situational development required, UW often presents unusual support challenges requiring imaginative and unorthodox logistical solutions. As a strategic policy option, logistical support cannot rely on merely one-echelon-up support arrangements; UW must be planned and resourced at every level of warfare by each required command level.

Geographic Combatant Command—Theater Strategic Level

4-40. SF units rely heavily upon regional or geographic combatant command theater support capabilities for support beyond their organic capabilities. For the most part, there is no habitual relationship of theater support units with their SF customers; early coordination and mutual support between all players during the development of a UW campaign is imperative. SF and other special operations elements deployed in support of a UW campaign are usually in isolated and austere locations requiring a mix of conventional, contracted, and ARSOF-unique capabilities and/or significant dependence on indigenous forces and neighboring allied nation support. More detailed information on Army logistics support to ARSOF is in ADRP 4-0, *Sustainment*. More detailed information on SF sustainment is in ATP 3-05.40, *Special Operations Sustainment*.

4-41. Sustainment planning must first consider the existing infrastructure. The Army retains responsibility for the sustainment of forces it allocates to a joint force. Logistics support covers supply, field services, maintenance, transportation, distribution, operational contract support, and general engineering support. Support providers include strategic-level providers, the generating force, and the operating forces within the Army. For example—
- GCCs have directive authority for logistics for their assigned forces ensuring effective execution of operational plans and economy-of-support forces, creating interdependent and efficient use of the different Service components' capabilities. The President or SecDef may extend directive authority for logistics to attached forces when transferring forces for GCCs' missions. Although Title 10 support falls under each individual military service, a combination of the different service capabilities within an area of responsibility (AOR) can create interdependent and efficient use of logistics sustainment capabilities. Sustainment of a joint force can be provided through a combination of executive agent, lead Service, and subordinate logistics command designations as per ADRP 4-0 and JP 4-0, *Joint Logistics*. Interagency coordination occurs between elements of DOD and USG agencies in order to achieve specific strategic objectives. Military operations must be coordinated with the activities of other agencies of the USG,

Chapter 4

- intergovernmental organizations, nongovernmental organizations, and activities of various host-nation agencies.
- The Army Service component command (ASCC) HQ is a theater of operations Army HQ with three functional roles—Service component; Title 10 Service, administration, and support; and, when directed by the GCC, the nucleus of a JTF HQ to command small contingency operations. The ASCC commander tailors his organization to provide or otherwise arrange for the required administration and support of deployed Army forces, including ARSOF. USASOC assigns an Army special operations forces liaison element (ALE) to the ASCC to coordinate logistics and to integrate support for deployed ARSOF. The ARSOF logistics planners identify the support requirements in the planning phase. The ASCC must also identify the logistics shortfalls for inclusion in the GCC's risk assessment in his AOR. If the ASCC cannot support ARSOF, the ASCC must raise the shortfall to the supported GCC for resolution.
- USASOC fulfills ASCC-like functions as a functional ASCC for SF, and is responsible for the manning and organization, training, equipping, administrating, and sustaining of ARSOF assigned to USSOCOM.
- U.S. Army Forces Command is an Army generating force that provides logistics forces to ARSOF in support of global operations through the Army force generation process.
- The TSOC is the joint special operations (SO) command through which the GCC normally exercises operational control (OPCON) of SOF within the AOR. The TSOC commander is also the permanent theater of operations joint force special operations component commander (JFSOCC). He commands the TSOC and is the principal SO advisor to the GCC. The TSOC is a subordinate command of a combatant command or a functional component command of another permanent joint command.
- Operating force support may consist of the theater support command, expeditionary support command, support brigade, combat sustainment support battalion, brigade support battalion, and brigade logistics support teams that are all integral supporting elements to theater operations. Theater support encompasses support JIIM forces.
- Host-nation support is civil and military assistance provided by a nation to foreign forces within its territory during peacetime, crises, or emergencies, or war based on agreements mutually concluded between nations.

Army Service Component Command and Theater Special Operations Command—Operational Level

4-42. The TSOC tasks missions to ARSOF. The TSOC works closely with the combatant command staff and the ASCC to articulate the ARSOF requirements. The GCC establishes priorities and allocates the available resources to ARSOF to accomplish each mission. The ASCC develops the AOR support plan, which includes sustainment of ARSOF by the AOR logistics organizations. The TSOC then monitors ARSOF sustainment.

4-43. The TSOC, ALE, and ARSOF support cell logisticians coordinate with the ASCC to develop plans and subsequent orders to implement directives the ASCC will issue to support the ARSOF assigned to the combatant command. The TSOC advises the ASCC commander on the appropriate command and support relationships for each ARSOF mission. The ALE keeps USASOC informed of the status of ASCC's supporting plans.

4-44. The 528th Sustainment Brigade (Special Operations) (Airborne) (SB[SO][A]) is USASOC's sustainment brigade and is deployable in support of ARSOF-led JSOTFs in austere environments. Its mission is to set the operational-level sustainment conditions to enable special operations including UW. The 528th SB(SO)(A) provides liaison to TSOCs and ASCCs, and may be tasked to coordinate directly with foreign vendors, U.S. embassies, and multinational forces. The 528th SB(SO)(A) provides a key link to the conventional force via embedded ARSOF liaison elements and ARSOF support operations teams in tactical and strategic conventional force sustainment structures.

4-45. The 528th SB(SO)(A) is unique when compared to other Army sustainment brigades in that it maintains global situational awareness of the ARSOF logistics support structure. The 528th SB(SO)(A) is designed to deploy small, modular teams. The 528th SB(SO)(A) can also serve as the senior logistics unit in the joint operations area. With the right augmentation and growth, it can establish theater-opening and intermediate staging base operations with tailored multifunctional Army sustainment enablers.

Special Forces Group Support Battalion—Tactical Level

4-46. When ASCC logistics support is unavailable or not established in the joint operations area, the SF group support battalion will be the primary common-user logistics provider. The group support battalion is a joint and multinational-capable organization in that it can accept, integrate, and employ augmentation of assets from other Services and nations. Augmentation may be necessary when the SOTFs are set up in undeveloped theaters without established Army theater opening, theater distribution, or area support structures.

4-47. The group support battalion provides rapidly deployable multifunctional logistics, force health service support, tactical and strategic communications, all-source intelligence, and multipurpose canine support. In developed theaters, the group support battalion synchronizes its support with the conventional force. A combined JSOTF may operate across entire countries or in multiple countries based on the operational construct. No single support battalion can cover this geographical dispersion. The group support battalion coordinates with conventional force sustainment brigades, the expeditionary sustainment command, and the theater sustainment command to enable an "area support" concept in support of each SOTF, advanced operational base, and SFODA. Area support enables SOF elements in the vicinity of conventional force bases to receive general sustainment support.

4-48. The group support battalion and SF battalion support companies may require Army logistics augmentation to provide logistics support during sustained operations or for a capability not organic to the SF group. This augmentation may be necessary when the SOTFs are set up in undeveloped theaters without established Army theater opening, theater distribution, or area support, when SOTF bases are not established at fixed facilities, or when a high percentage of SF operational detachments are committed simultaneously.

4-49. The headquarters and headquarters detachment of the group support battalion provides organic battalion-level administrative and supply support for all assigned and attached personnel and coordinates external support for the SF battalions through the support operations section. The support operations section within the group support battalion is the hub of multifunctional logistic operations in support of the SF group. The support operations section—

- Conducts continuous logistics preparation of the battlefield.
- Develops and synchronizes the overall concept of support.
- Plans and coordinates for ground and air resupply.
- Plans for landing zones in the vicinity of the SOTF.
- Develops logistics synchronization matrixes.

More detailed information concerning the SF group and below logistic information can be found in FM 3-18, *Special Forces Operations*.

LEGAL CONSIDERATIONS

4-50. UW often inhabits the gray area of political engagement where it blends into warfare. It is crucial that Soldiers planning and executing UW pay careful attention to legal considerations and respect for authorities.

EXECUTION AUTHORITIES AND SENSITIVITIES

4-51. Military operations require an instrument authorizing their execution. The senior policy document which describes the activities and operations of UW is USSOCOM Directive 525-89. This directive conforms to U.S. statutes, the Law of Land Warfare, and rules of engagement issued by competent authority. The complexity associated with a UW campaign (because of its varied preparatory activities, range of operational environments, operational modes, and extended nature) may require a multitude of authoritative instruments, each with its own unique constraints, restrictions, limitations, content, process, and office wherein lies legal authority to approve its execution.

4-52. In order to be successful, UW campaigns depend on the secrecy of the campaign's objectives, participants and relationships, locations, capabilities, timing and synchronization, and the tactics, techniques and procedures to be used. The intent is to protect sensitive information, but not impede the planning and execution of UW campaigns.

COVERT ACTION

4-53. No agency except the Central Intelligence Agency—or the Armed Forces in time of war—may conduct any covert action activity unless the President determines that another agency is more likely to achieve a particular objective. 50 USC Section 413, *General Congressional Oversight Provisions*, defines covert action as "activity or activities of the USG to influence political, economic, or military conditions abroad, where it is intended that the role of the USG will not be apparent or acknowledged publicly, but does not include—

- Activities, the primary purpose of which are to acquire intelligence, traditional counterintelligence activities, traditional activities to improve or maintain the operational security of USG programs, or administrative activities.
- Traditional diplomatic or military activities or routine support to such activities.
- Traditional law enforcement activities conducted by USG law enforcement agencies, or routine support to such activities.
- Activities to provide routine support to the overt activities (other than those described above) of other USG agencies abroad.

4-54. Title 50 USC 413(b) provides Presidential authority and limitations in regard to the issuance of findings authorizing covert action and documentation and oversight requirements. The President may not authorize the conduct of a covert action by departments, agencies, or entities of the USG unless he determines such an action is necessary to support identifiable foreign policy objectives of the USG and is important to the national security of the United States, which determination shall set forth in a finding that shall meet each of the following conditions:

- Each finding shall be reduced to a written finding as soon as possible, but in no event more than 48 hours after the decision is made.
- Unless immediate action by the USG is required, and time does not permit the preparation of a written finding, a finding may not authorize or sanction a covert action, or any aspect of any such action, which has already occurred.
- Each finding shall specify each department, agency, or entities of the USG authorized to fund or otherwise participate in any significant way in such action. Any employee, contractor, or contract agent of a department, agency, or entity of the USG other than the Central Intelligence Agency directed to participate in any way in a covert action shall be subject either to the policies and regulations of the Central Intelligence Agency, or to written policies or regulations adopted by such department, agency, or entity, to govern such participation.
- Each finding shall specify whether it is contemplated that any third party which is not an element of, or a contractor or contract agent of, the USG, or is not otherwise subject to USG policies and regulations, will be used to fund or otherwise participate in any significant way in the covert action concerned, or be used to undertake the covert action concerned on behalf of the United States.
- A finding may not authorize any action that would violate the Constitution or any U.S. statute.

4-55. 50 USC 413(b) requires the congressional intelligence committees to be kept fully and currently informed of all covert actions. Any covert action finding shall be reported to the committees as soon as possible after such approval, and before the initiation of the covert action authorized by the finding.

PREPARATION OF THE ENVIRONMENT

4-56. The broad nature of PE activities precludes identification of specific authorization requirements, but common to all is the requirement to obtain authorization as prescribed within applicable orders, laws, and regulations. Coordination procedures for PE vary widely, but common to all is the TSOC's responsibility for the development, coordination, and implementation of PE activities with the respective country teams, regional interagency, and GCC counterparts. Detailed information on PE is located within USSOCOM Directives 525-16 and 525-5.

4-57. Military intelligence operations and activities are authorized by 10 USC. Such activities include, but are not limited to—

- Human intelligence.
- Counterintelligence.
- Signals intelligence.
- Tagging, tracking, and locating operations.
- Close-in persistent surveillance.
- Intrusive reconnaissance and surveillance.
- Measurement and signatures intelligence.
- Several classified activities or operations that support PE.

Such 10 USC military intelligence operations and activities are executed under the GCCs' 10 USC authority.

4-58. The DOD may also conduct intelligence operations and activities under authority of 50 USC in support of national intelligence requirements. Such operations are conducted under authority granted to the SecDef under 50 USC. Such operations are approved for execution by the SecDef following GCC or USSOCOM commander endorsement and favorable Director of Central Intelligence directive coordination by Director, Central Intelligence Agency; Director, National Security Agency; and pertinent chiefs of mission, chiefs of station, and the Undersecretary of Defense for Intelligence.

This page intentionally left blank.

Chapter 5
Campaign Planning

To shape the future force, we must grow leaders who can truly out-think and out-innovate adversaries while gaining trust, understanding, and cooperation from our partners in an ever-more complex and dynamic environment. The enduring challenges we face and the whole-of-nation approaches they require demand leaders that have the qualities of flexibility, agility, and adaptability, and the ability to build unique teams of teams to accomplish missions.

National Military Strategy, 2011

A campaign is a series of related major operations aimed at accomplishing strategic and operational objectives within a given time and space. Planning for a campaign is appropriate when the contemplated simultaneous or sequential military operations and activities exceed the scope of a single major operation. While intended primarily to guide the use of military power, joint operational plans for campaigns should consider all instruments of national power and how their integrated and/or coordinated efforts work to attain national strategic objectives. This coordinated approach is particularly important for campaign planning since, in many instances, other USG agencies have the lead role for shaping activities, which DOD efforts must complement and support.

Note: This chapter provides fundamental but generalized and generic considerations for UW campaign planning. The chapter does not represent an exhaustive coverage of all relevant planning considerations. It is intended that commanders and planners will use this publication and chapter in conjunction with the relevant theater campaign plan; JP 5-0, *Joint Operation Planning*; ADRP 5-0, *The Operations Process*; and the U.S. Army War College *Campaign Planning Handbook*.

UNCONVENTIONAL WARFARE PLANNING CONSIDERATIONS

5-1. The potential of UW to be an effective national strategic policy option is based in part on effective understanding and articulation of how UW fits into/relates to generic joint campaign planning. Campaign planning considerations which are unique, subtly different, and/or particularly relevant to UW include the following:
- UW is a national strategic policy option.
- Any campaign planning effort has a potential UW component.
- UW is conducted to coerce and disrupt some opponents and, in relatively rare cases, to overthrow others. These three purposes are different and imply different levels of commitment, scale, and sensitivity.
- UW is basically conducted in one of two ways: either as one of many subordinate lines of effort supporting a larger conventional campaign, or as a main effort supported by other JIIM partners.
- In UW, the emphasis is on high-value strategic and operational return on investment rather than on tactical mass or firepower.
- UW emphasizes the indirect approach.

Chapter 5

- Although UW is a U.S. initiative, the emphasis is on organizing, force-multiplying, and leveraging indigenous partners. Theoretically, U.S. forces can conduct UW successfully without ever firing a round.
- In UW, the involvement of other JIIM partners is essential. JIIM involvement is not an adjunct to military planning. The more UW is the campaign main effort, and the more UW is conducted during peacetime, the more central the JIIM involvement will be. UW planners make maximum appropriate use of JIIM enablers.
- Although the conduct of UW is dependent on requisite legal authorities, UW design and execution is flexible and does not always follow doctrinal and procedural formats.
- Many activities inherent to UW may be conducted by other agencies outside of and prior to formal UW campaign planning. The effort technically and legally only becomes "UW" upon issuance of a presidential finding.
- In some cases, UW may take years or decades to develop. Therefore, national and theater strategic foresight and long-term commitment are critical.
- UW is human-centric. On-the-ground expertise with micro-conditions in the human domain which provides accurate ground truth early in the planning process is imperative.
- Each instance of UW is unique, and takes full account of the specific existing geopolitical and demographic context. Therefore, the details of such planning will usually by developed by joint planning groups (JPGs) of a TSOC subordinate to a GCC.

Seven Phases of U.S. Army Special Forces Unconventional Warfare—Key Considerations

The seven-phase UW plan is the classic U.S. Army SF template for the planning and execution of UW. The seven phases are a conceptual guideline from which planners may adjust the details as necessary for the specific conditions of any given UW effort. It is imperative that commanders and planners remember that each UW effort will be different. Not all phases may apply, phases may not progress in the prescribed linear order, phases may regress, and/or separate elements executing UW activities may operate in different phases within the larger context of the GCC's overall campaign. Like all doctrinal constructs, the seven phases should guide and inform commanders and planners, not constrain them. Some key considerations of the seven UW phases are as follows:

- *Phase I—Preparation.* UW is an act of war. It does not begin until there is explicit Presidential or SecDef authority to execute it. Many non-UW activities may occur prior to UW phase I, which may be retroactively included in a UW campaign after authority is granted to execute. Such activities will be part of the Joint Operation Planning and Execution System (JOPES) Phase 0 (Shape) of one or more GCC campaign plans. Therefore, PE is not in itself "UW preparation." UW Phase 1 preparation activities may be conducted by other government agencies.
- *Phase II—Initial Contact.* Like UW Phase 1, UW Phase 2 may be done by other agencies prior to approval of a specific UW campaign plan. UW Phase 2 is largely an intelligence fact-finding mission.
- *Phase III—Infiltration.* The key task of UW Phase 3 is successful linkup with insurgent/resistance forces. At that point, infiltration is complete. Insertion is merely the method of entry into denied territory.
- *Phase IV—Organization.* The key idea is developing an agreed upon plan with counterparts. It takes time to understand what makes sense and develop that relationship.

- **Phase V—Buildup.** Plans agreed to in UW Phase 4 are implemented in UW Phase 5. The amount of effort required to conduct organization building will be based on the insurgent/resistance organization that may already exist. Organization could be time consuming and painstaking or friendly forces could fall in on well-established and robust organizations. When insurgent/resistance organizations are fragile and hard-pressed by effective counterinsurgent forces, growing too big too fast is worse than being too small. Premature and unwanted exposure of the insurgent/resistance organization can be deadly. This represents the most dangerous time for an insurgency and highlights why organizational planning is so critical to get right.
- **Phase VI—Employment.** This will be a series of many events—both lethal and nonlethal—throughout UW Phases 4–7. In other words, UW Phase 5 does not come to a complete stop and only then does UW Phase 6 begin. Nevertheless, this may represent a planned surge or maximum growth of organization effectiveness synchronized with planned future operations.
- **Phase VII—Transition.** When UW ends in overthrow of a state or liberation of occupied territory and leads to a new government, UW Phase 7 includes those activities contributing to the promotion of the new government's legitimacy. In such cases, the UW effort transitions to foreign internal defense at some point. However, not all UW efforts will result in victory or liberation. This is true when the object of UW is relatively limited coercion and or disruption of the opponent. In such cases, transition may be withdrawal of U.S. forces, modification in the levels of support, and/or turning active insurgent or resistance forces dormant.

5-2. While the U.S. Army SF seven phases of UW are valuable in planning UW operations, SOF commanders must articulate their vision to accomplish national objectives as part of a tailored UW campaign. SOF must translate the seven SF doctrinal phases of UW into a campaign construct that uses phases and terms that are meaningful at the GCC level and higher. Commanders should outline their objectives, desired effects, lines of activity, decision points, required authorities, and necessary resources across all phases of the campaign. The UW campaign plan needs to explain how SOF will build or strengthen a resistance organization (underground, auxiliary and guerrilla force), win the support of the population, and defeat the adversary regimes' instruments of power. The inclusion of regional partners who can contribute forward staging areas and local knowledge, for example, is of paramount importance. Ultimately, commanders need to clearly describe to a variety of stakeholders—including civilian officials—how military actions will accomplish political ends. Commanders should tailor the UW campaign plan to the specific character and circumstances of the regime and conditions in the operational environment.

JOINT OPERATION PLANNING PROCESS

5-3. The joint operation planning process (JOPP) provides a structured process to formulate a mission, develop appropriate COAs to accomplish the mission, and coordinate and integrate the details of a plan to execute the selected COA. The JOPP is a seven-step process.

STEP #1: INITIATION

5-4. Upon receiving strategic guidance from higher HQ, or as directed by the commander, the staff begins the JOPP by initiating planning. The commander uses a design methodology to refine his understanding of the situation and to develop his operational approach. Based on this understanding, the commander issues initial planning guidance to the JPG. The JPG receives and reviews a series of documents that outline the situation and background of the planning problem. These will include—
- A theater campaign plan which provides details on current activities in the GCC.
- A planning order which provides authorities to execute specific activities.
- Regional and country-specific action plans.

Chapter 5

5-5. JPGs must understand both the time frame in which desired campaign end states are to be achieved and the time schedule for planning itself. They determine who should be involved in the various aspects of planning, to include appropriate multinational partners and representatives of other USG agencies. They should also consider who the other interested parties are that may help formulate the plan, or should be brought in to achieve their buy-in of the plan.

5-6. Theater campaign plans serve as the base plan for UW campaign planning and, unless otherwise directed, are considered contingency plans to the theater campaign plan. However, since UW is typically a long-term sensitive and complex JIIM effort, conditions of the UW plan may cause such a shift in the environment that a revised theater strategy and theater campaign plan will be required.

STEP #2: MISSION ANALYSIS

5-7. The staff analyzes the mission to provide a recommended mission statement to the commander, and to provide detailed analysis to inform the commander's analysis of the operational environment and the problem and help him shape an operational approach. The key products of the mission analysis are running estimates—continuous assessments of current and future operations from the perspective of supportability and which inform the commander's COA development.

Analyze the Higher Headquarters' Planning Directives and Strategic Guidance

5-8. It is absolutely critical that the GCC is able to accurately and effectively articulate the national strategic utility of conducting a UW campaign in his theater. The quality of the commander's articulation is directly informed by the quality of strategic analysis by the TSOC and JPG. JPG staff work is guided by the commander's designed framework, and seeks initially to determine the national strategic end state, the theater strategic/military end state, and strategic objectives:

- The national strategic end state describes the President's political, informational, economic, and military vision for the region or theater once operations conclude. An example might be "An economically viable and stable Country X, without the capability to coerce its neighbors." National strategic end states derive from President/SecDef guidance that is often vague. More often than not, senior military leaders will assist the President/SecDef in developing and articulating that end state. It is predominantly in this senior leader interaction where the articulation of UW's strategic utility is achieved. The UW campaign planning staff must properly inform and prepare the commander for this interaction.
- The military end state is a subset of the national strategic end state discussed above and generally describes the military conditions necessary to achieve the national strategic end state. The JPG will develop the military end state after analyzing the tasks required by strategic direction. An example of a military end state is "Country X is unable to project military power against its neighbors."
- Strategic objectives clarify and expand upon the end state by defining the decisive goals to achieve in order to assure U.S. policy. Objectives prescribe friendly goals. The following are examples of strategic objectives:
    - "Indigenous forces supported by the United States and its allies have established a secure and stable government in Country X before U.S. forces depart."
    - "X ceases support to regional terrorism."

5-9. The end state defines achievement of the commander's campaign plan objectives and seeks to answer the question, "How does the U.S. strategic leadership want the operational environment to function at the conclusion of the campaign?" Strategic objectives normally answer the "what" of a campaign plan and seek to answer the question, "What needs to be done to achieve the end state?"

5-10. Answering these questions is what makes mission analysis different at this level when compared to the operational and tactical levels. There is no clear, definitive guidance in one location. There is no higher order from which to simply cut and paste into the emerging plan's operation order. Instead, the National Security Strategy (NSS), the National Defense Strategy (NDS), the National Military Strategy (NMS), Presidential Policy Directives, SecDef and Presidential speeches, and verbal guidance all provide direction—often disparate and contradictory—to help define an end state and corresponding objectives. U.S. Army SF maintain the regional expertise, interagency experience, and UW campaign planning skills to effectively distill these end states and objectives and must be included in such planning.

5-11. The role of strategic communication guidance must also be considered starting at this initial and critical planning step. This guidance often provides clarity to other guidance, but should also be considered a source of limitations. It specifies how the USG will engage key audiences to create, strengthen, and/or preserve conditions favorable to accomplish national policy objectives. The guidance may also describe the coordination of programs to inform and influence key audiences and provide limitations on what and what not to say and do in planning and executing the campaign. The strategic communication guidance shapes not only the commander's strategic communication guidance, but shapes the whole mission, and it certainly may provide some explicit and implicit limitations that must be considered.

Determine Known Facts and Assumptions

5-12. Facts are the major pieces of information known to be true and that are pertinent to the planning effort. Successful UW campaign planning requires broad and detailed understanding of the operational environment and all pertinent geostrategic factors, both great and small, must be synthesized during mission analysis. The factual emphasis in UW planning is on human political factors of organization and overall potential for resistance. Facts characterize the known reality of the operational environment within which the strategic end state will be reached. Consideration of facts and their interrelationship enhance mission analysis. The JPG should leverage the strategic estimate as a useful means to organize and consider geostrategic factors in an attempt to gain a better understanding of their impact and interrelationships. This analysis will include analysis of the PMESII-PT variables, which include the physical environment (topography, hydrography, climate, weather, and demographics) and time (the effect timing aspects have on the operational environment and on the campaign). The key is to determine potential effects of all these variables on possible operations of friendly, neutral, adversary, and enemy military forces and other instruments of power. Appendix F contains a checklist of considerations to establish and organize facts.

5-13. The staff develops assumptions in order to continue the planning process in the absence of facts. Assumptions are placeholders to fill knowledge gaps, but they play a crucial role in planning and must be held to a minimum throughout planning. These assumptions require constant revalidation and reassessment. Facts may replace them as more information is available. A planning assumption must be logical, realistic, and essential to continuing the analysis and planning. It is logical and realistic if there is sufficient evidence to suggest that it will become a fact. It is essential if required for planning to continue. Assumptions should also be clear and precise. Normally, the higher the command echelon, the more initial assumptions exist. Incorrect or risky assumptions may partially or completely invalidate the entire plan. In UW, the quality of assumptions for UW planning is directly correlated to genuine expertise regarding the human population in the target JSOA. This highlights why U.S. Army SF, DOS, and OGA representatives with long-established expertise in the target JSOA must constitute the primary planners of a UW campaign plan JPG. Examples of theater-level assumptions include—

- Political:
    - Countries A and B will allow overflight, basing, and host-nation support, to include covert and clandestine operations from their territory.
    - Country C will support Country E with air and naval forces, and will provide political, intelligence, communications, and logistical support.
    - Country C land force intervention in support of Country E up to light infantry regiment strength is possible.
    - Country D will remain neutral.
    - Exposure of proven U.S. involvement in Country E will provoke international censure and sanction.
    - International censure and sanction of U.S. UW campaign activities will result in mission failure.

- Forces:
  - Two SF line battalions and elements of one support battalion will be available for this campaign.
  - Two naval special warfare platoons, one sea-air-land team delivery vehicle platoon, and two Marine special operations companies will be available for this campaign.
  - Elements of one Air Force special operations squadron will be available in-theater.
- Timeline: Upon receipt of execute order PE time frame not later than 18 months prior to D-day.
- Enemy:
  - Country E's counterinsurgency response will result in deliberate targeting of civilian population.
  - Country E will use chemical weapons against suspected coalition forces locations within the boundaries of Country E.

Determine and Analyze Operational Limitations

5-14. Limitations are the restrictions placed on the commander's freedom of action. They may be part of strategic direction or stem from regional or international considerations or relationships. Limiting factors are generally categorized as constraints or restraints. Constraints are tasks that the higher authority requires subordinates to perform. Constraints are must-do actions. Restraints are things the higher authority prohibits a subordinate commander or force from doing. Restraints are must-not-do actions.

5-15. UW campaigns are especially susceptible to limitations. UW campaigns are characterized by clandestine and covert operations with indigenous partners in denied territory and across international borders over an extended time period. The politically sensitive nature of many UW activities, and the reluctance of states to have their involvement exposed, will require carefully crafted rules of engagement, specifically crafted and strictly monitored authorities, and imaginative application of indirect methods to achieve objectives. UW campaigning is not typically conducted by straightforward application of U.S. power or by automatic processes. Indeed, much of UW's strategic utility is derived from its ability to meet U.S. objectives in ways that avoid the limitations placed on overt use of large conventional forces for obvious objectives. UW is primarily about leveraging and empowering local actors to conduct operations in ways that exploit, circumvent, or obviate limitations.

Determine Specified, Implied, and Essential Tasks

5-16. UW campaign planning develops national strategic and theater strategic options to meet strategic end states. Specified tasks for such planning will be necessarily broad and will likely require using many elements of national power and the action of several elements of the joint force. Specified tasks do not specify actions by components or forces. Examples of specified tasks to a combatant command include—
- Deter Country E from coercing its neighbors.
- Stop Country E's aggression against its own population.
- Reduce Country E's ability to traffic in WMD materials and narcotics, to conduct human trafficking, and to support international terrorism.
- Disrupt Country E's regime by supporting the indigenous populations' efforts to establish a viable security structure and political challenge to the Country E regime.
- Remove Country X's regime.
- Prevent intervention by Country C.

5-17. After identifying specified tasks, the staff identifies additional major tasks necessary to accomplish the assigned mission. These additional major tasks are implied tasks. These are tasks that the joint force must do in order to accomplish the higher echelon's specified tasks. Tasks that are inherent responsibilities, such as deploy, conduct reconnaissance, and sustain, are not implied tasks unless successful execution requires coordination with or support of other commanders. Examples of implied tasks are—
- Build and maintain or enhance existing indigenous forces for a UW campaign.
- Conduct PE in Country E.

Campaign Planning

- Develop indigenous, expatriate, and multinational political entities to oppose Country E regime.
- Destroy Country E's counterinsurgency constabulary effectiveness.
- Provide assistance to newly constituted indigenous regime in the wake of current regime removal.

5-18. Essential tasks derive from the list of specified and implied tasks and are those tasks that the joint force must conduct in order to accomplish the mission successfully. These are the tasks that appear in the mission statement.

Determine Termination Criteria, Military End State, Objectives, and Initial Effects

5-19. Campaign design and planning translates national or theater strategic objectives into action by integrating end states, objectives, effects, and tasks among all components of the command. The commander and staff determine the appropriate termination criteria for military activity that will enable achievement of the national strategic end state. These criteria describe the conditions in-theater that must be met before conclusion of the campaign or operation, or before transition of the campaign to a supporting campaign to enable other elements of power to achieve the national end state.

5-20. The commander and staff must then translate those criteria into a concise statement of the military end state, develop a set of objectives that will lead to achievement of the military (or theater) end state, and begin analysis to determine the effects (specific conditions) required to achieve the objectives (Figure 5-1). As the strategic aims shift, so too must consideration of termination criteria and operational objectives. The commander may provide significant change to his guidance as a result of his reassessment of the operational environment, the associated problem definition, and the resulting adjusted operational approach. This significant change is reframing and will almost certainly result in adjusted planning guidance.

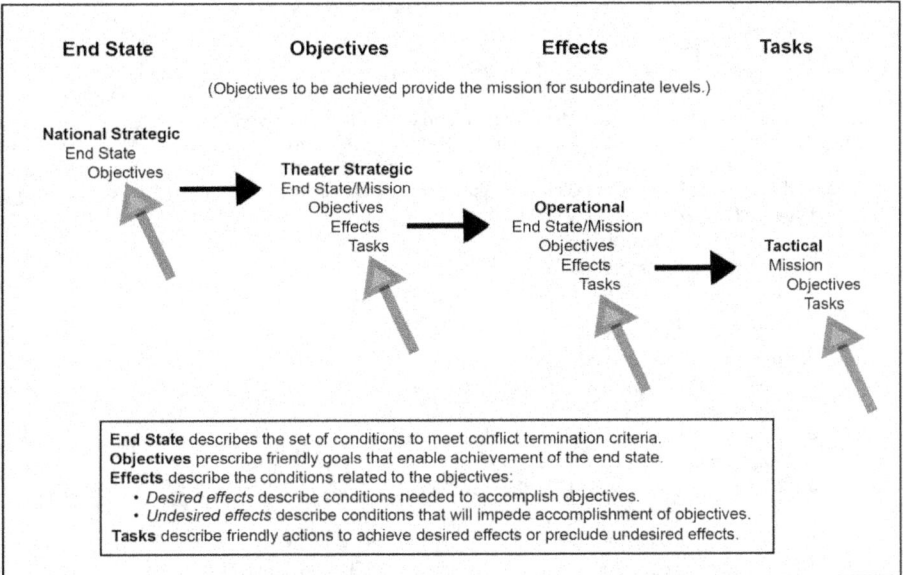

Figure 5-1. Relationship of end state (mission), objectives, effects, and tasks

Chapter 5

Termination Criteria

5-21. Termination occurs at that point when the military end state is achieved and the achieved advantages can be preserved. Termination criteria describe the specified standards to meet before conclusion of a joint operation. When addressing conflict termination, commanders and their staffs must consider a wide variety of operational issues, to include disengagement, force protection, transition to post-conflict operations, reconstitution, and redeployment. Termination criteria may change as the campaign progresses, and commanders and staffs must anticipate such changes. This requires continuous coordination with the civilian national leadership, the leadership of other USG agencies, and other partners. Some examples of termination criteria follow (note the description of condition, not action):

- Country E no longer presents a coercive threat to neighboring states.
- Country E has enacted negotiated terms for power sharing with indigenous population.
- Country E is no longer a reliable transshipment point for WMD materials, narcotics, terrorists, or human traffickers.

Military End State

5-22. The military end state describes the military conditions to satisfy the objectives of the national strategic end state. Often, the military end state is achievable before the national strategic end state, signifying when the President no longer requires the military as the primary element of national power required to achieve the remaining objectives of the national strategic end state. However, achievement of the military end state will likely not end military involvement as a supporting effort. An example of a military end state is—

- Country E does not possess sufficient offensive military power to threaten neighboring states.
- Country E no longer possesses sufficient police capability to repress its own indigenous populations.

Campaign Objectives

5-23. Objectives are the clearly defined, decisive, and attainable goals toward which joint actions focus to accomplish the military end state. Military objectives are one of the most important considerations in campaign and operational design. They define the role of military forces in the larger context of national strategic objectives, and are nested within national strategic objectives. They specify what to accomplish and provide the basis to describe campaign effects. Objectives may change throughout the campaign, either because the end state changes or because the operational environment changed to such an extent that there are better objectives to achieve the end state. Primary considerations for an objective include the following:

- Establishes a single desired result (goal).
- Links to higher objectives directly or indirectly.
- Is prescriptive, specific, and unambiguous.
- Does not infer ways and/or means (not written as a task).

5-24. Examples of theater strategic military objectives that support the above military end state might be—

- Country E's offensive military capabilities are reduced to prevent attack of neighboring countries.
- Country E no longer supports regional insurgent or terrorist groups that threaten stability in neighboring countries.
- Country E possesses only defensive capabilities and is integrated into regional cooperative defense arrangements.
- Country E's counterinsurgency constabulary is combat ineffective.
- Indigenous populations' participation and capability in Country E's security apparatus is sufficient to protect their interests.

Campaign Effects

5-25. An effect is a physical and/or behavioral state of a system that results from an action, a set of actions, or another effect. Effects are an essential element of campaign design. The commander will provide his vision on the desired and undesired effects to achieve through unified action. In turn, the staff and subordinate commands will use these effects to analyze the tasks and resources required to create the effects desired and preclude the undesired effects. Effects may be expressed in two ways:

- *Desired effects*: How does one want the operational environment to behave when the end state is reached, or at particular points of the campaign en route to the end state?
- *Undesired effects*: What are the behaviors and conditions in the operational environment that must be avoided during the campaign?

5-26. Objectives describe what the commander wants to do, effects describe what conditions must exist for accomplishment of objectives, and tasks describe what actions must be taken to achieve those effects.

5-27. Other agencies and partners will use the desired effects to help them visualize their activities to support the military activity if it is the primary means of achieving objectives, or how they will be supported by military activity if they provide the key means to achieve national strategic objectives. Thus, effects may be a prime means to bridge military and interagency understanding by describing how the operational environment should behave to show policy achievement.

5-28. Joint doctrine has no specific convention for writing effects, but there are five primary considerations:

- They should link directly to one or more objectives.
- They should be distinguishable from the objective they support for success.
- They should be stated as conditions of the operational environment.
- They should be measurable.
- They should not specify ways and means for accomplishment.

5-29. During mission analysis, the staff develops broad, overarching effects for the commander's consideration and later use. Examples of campaign effects are—

- Country E's military forces are not conducting cross-border offensive military operations.
- Country C is not supporting Country E in destabilizing the region.
- Regional terrorist networks are not influencing regional governments and populations.
- Local governments are effectively administrating and providing support to the populace.

Write the Initial Mission Statement

5-30. After identifying the essential tasks, and with the context of the relationship of those tasks to the achievement of the national end state and military end state, the staff normally develops a derived mission statement using the format of who, what, when, where, and why. This statement should be a direct, brief, and effective articulation of the essential tasks and purpose for military operations. This mission is also critical in that the commander of the next-higher echelon will approve it, or its key elements, and in the case of a GCC, the SecDef and the President will most likely adopt the key elements of the mission statement as they orchestrate unified action and articulate the rationale for military operations to potential coalition partners. A mission statement might look like this: "On order, CJSOTF-X, in partnership with the Free Forces Army, will disrupt Country E capability in JSOA X in order to compel the regime to abandon its support to terrorism in Country Y."

Determine Mission Success Criteria

5-31. Mission success criteria describe the standards for determining mission accomplishment. These criteria help the commander determine if and when to move to the next phase or next major operation. This set of criteria becomes the basis for the assessment plan. Mission criteria should be set for the campaign overall, but also for each subordinate phase or operation. As these mission criteria should be echeloned and nested, they will necessarily be different for each level of command, focused on accomplishment of that command's mission.

Develop Decisive Points and Lines of Operation

5-32. Once the mission has been derived, and the commander and staff have a good appreciation for how their mission fits into accomplishment of the overall national strategic end state, the staff relooks the critical factors in the context of that mission. Planners need to understand the sources of the enemy regime's power so that SOF can devise a strategy that maximizes pressure at key points.

5-33. There are two main streams of thinking that lead to development of decisive points and lines of operation/lines of effort. The first is through the design methodology of describing the tensions and opportunities presented by the various actors' desired conditions and the likely activities that will enable achievement of those conditions. The second is through PMESII-PT analysis and center-of-gravity analysis to determine the critical vulnerabilities that can be exploited to defeat the adversary and must be protected to sustain friendly efforts.

Refine PMESII-PT Analysis

5-34. The value of PMESII-PT analysis is to identify the key actors and relationships, as well as those key physical and nonphysical aspects of the environment. (Appendix F includes a detailed analysis format.) Below is a representative sample of inquiries pertinent to UW that should be included in a PMESII-PT analysis:

- Political:
    - Who is who" in the targeted regime? Who are the moderates? Who are the hardliners?
    - How can the moderates be strengthened? How can the hardliners be discredited?
    - How can the regime elites be split from the common people?
    - What are the resistance group's organizational structure, leadership dynamics, and sources of legitimacy?
    - What role could/should SOF play in influencing these dynamics?
    - What are the regime's relations with neighboring countries? What are its key international relationships?
- Military:
    - What are the enemy's land, air, maritime, and space capabilities?
    - Which units are most loyal or least loyal to the regime?
    - What is the military potential of the resistance group? How can the means be developed to gain access to them?
- Economic:
    - What are the characteristics of the target regime's economy? How is it sustained?
    - What is the impact of economic sanctions on the regime? What is the impact on the populace?
- Social:
    - What are the detailed demographics on all relevant populations in the AO? What are the key dynamics between population groups?
    - What are the specific demographics characterizing the resistance group (tribal, ethnic, social, political, and so on)?
    - How likely are indigenous population groups to support or resist U.S. UW activities?
- Informational:
    - What are the enemy's propaganda, counter-space, countersignal, and biometric capabilities?
    - How important is the Internet to the society? How widespread is access to the Internet?
    - How effective are regime cyber warfare capabilities to protect its own assets? How effective are regime censors in denying information to the populace?

Campaign Planning

- Infrastructure:
  - What are the physical characteristics of the AO (topography, roads/trails network, mountain passes, population centers, hydrographic characteristics of coastal areas and rivers, and so on)?
  - What food, shelter, transportation, intelligence, and so on is the indigenous population able and willing to provide the resistance?
- Physical environment:
  - What are the physical characteristics of the AO (topography, roads/trails network, mountain passes, population centers, hydrographic characteristics of coastal areas and rivers, and so on)?
  - What natural hazards may impede the target regime's ability to counter UW activities?
  - What climate or weather conditions in this AO are favorable to UW?
- Time:
  - Does the indigenous population have a different cultural perception of time?
  - Will popular support for extended UW operations diminish over time?
  - At what times are people likely to congest roads or conduct activities that provide cover for UW actions?

Conduct Center-of-Gravity Analysis

5-35. The analysis of relevant centers of gravity is a key step in the campaign design process. The purpose is to provide a base of understanding of friendly and adversary (and possibly neutral) systems which, in combination with the PMESII-PT analysis, enable development of decisive points that will contribute to achievement of campaign objectives. Center-of-gravity analysis must include the friendly and enemy strategic and cultural perspectives.

5-36. The center of gravity always links to the mission and its supporting objectives; therefore, as the mission changes, the center of all strength and power to accomplish or oppose these objectives may change as well. Planners should strive to identify only one center of gravity at any level of war and at any given time in the campaign or the term will lose its meaning and usefulness. Assuming the strategic end state or objectives do not change, normally the strategic center of gravity will not change during the campaign. However, objectives are likely to change during a campaign as one major operation transitions to another. In this event, the operational center of gravity may change.

> **Example**
> In prehostilities during a UW campaign, the United States may seek to politically isolate an enemy regime whose initial center of gravity is diplomatic support that permits it to act with relative impunity. However, once the United States has achieved some measure of international isolation on the regime, and has initiated support for indigenous resistance groups within the AO, the enemy's new center of gravity may be the strength of its internal security and repression apparatus. This would then become the new focus of U.S. operations.

5-37. *Center of gravity* is the set of characteristics, capabilities, and sources of power from which a system derives its moral or physical strength, freedom of action, and will to act. *Critical capabilities* are those capabilities deemed to be crucial enablers for a center of gravity to function as such and are essential to achieve the desired objectives. *Critical requirements* are essential conditions, resources, and means for a critical capability to be fully operational. These may be such things as communications nodes, bandwidth, or financial resources to buy weapons or influence. *Critical vulnerabilities* are those aspects or components of critical requirements that are deficient, or vulnerable to direct or indirect attack in a manner that achieves decisive or significant results.

Chapter 5

Determine Decisive Points

5-38. A decisive point is a geographic place, specific key event, system, or function that, when acted upon, allows commanders to gain a marked advantage over an adversary or contribute materially to achieving a desired effect, thus greatly influencing the outcome of an action. Developing decisive points orients on the key vulnerabilities or other critical factors identified through center-of-gravity analysis. Ideally, commanders design campaigns that attack adversary vulnerabilities at decisive points so that the results they achieve are disproportionate to the resources applied. Commanders and their staffs must determine and prioritize which vulnerabilities or capabilities, or key events, offer the best opportunity to achieve the effects on the operational environment that are needed to accomplish objectives. Some potential decisive points might include—

- In-theater ports, airfields, rail lines, or roads needed for support of UW operations are coordinated in neighboring countries.
- Pilot teams have infiltrated the operational area, linked up with, and assessed resistance potential of indigenous groups.
- Human infrastructure has been established in support of future UW operations.
- Infiltration routes have been established.
- Indigenous resistance groups have sustained success in guerrilla operations against the regime.
- Country E's counterinsurgency operations are significantly degraded.
- International community begins to recognize resistance movement as a legitimate political actor in Country E.

Develop Lines of Operation/Lines of Effort

5-39. Commanders may describe the operation along lines of operation or lines of effort, or both. Lines of operation or lines of effort offer a good framework for the commander to describe his visualization of a campaign. The staff refines lines of operations/lines of effort as it conducts COA development, and throughout execution of the campaign, to ensure coordinated, synchronized joint action.

5-40. A line of operation is a logical line that connects actions or nodes and/or decisive points related in time and purpose with an objective. A line of operation is also a physical line that defines the interior or exterior orientation of the force in relation to the enemy or that connects actions on nodes and/or decisive points related in time and space to an objective. A line of effort links multiple tasks and missions using the logic of purpose—cause and effect—to focus efforts toward establishing operational and strategic conditions.

5-41. Planners may use both lines of operations and lines of effort to build their broad concept. Lines of operation portray the more traditional links among objectives, decisive points, and centers of gravity. A line of effort, however, helps planners link multiple tasks with goals, objectives, and end state conditions. Combining lines of operation and lines of effort allows planners to include nonmilitary activities in their broad concept. This combination helps commanders incorporate stability or civil support tasks that, when accomplished, help set end state conditions of the operation. Figure 5-2, page 5-13, shows a sample line of operation for UW.

5-42. Developing lines of effort requires creative analysis and the ability to envision how potentially decisive events throughout the campaign link together. In developing and portraying lines of effort, the staff must—

- Understand and portray the critical initial conditions in the operational environment.
- Understand and portray the desired theater conditions with some sense for timing of those conditions.
- Understand and portray the theater objectives for the campaign (phased over time as needed) to achieve the desired theater conditions.
- Array decisive events (actions, functions, and so on) and/or locations for adversary and friendly efforts.
- Examine the decisive points and group or connect them into patterns or unifying factors.
- Collect and describe the pattern or unifying factors into lines of effort or unifying themes that run throughout the campaign. Figure 5-3, page 5-14, shows sample lines of effort for UW.

Campaign Planning

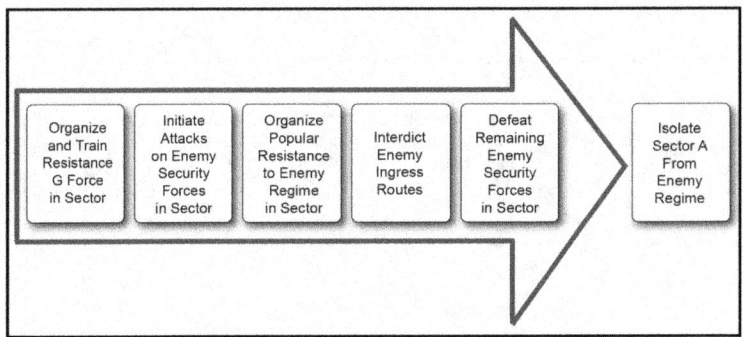

Figure 5-2. Sample line of operation for unconventional warfare

Review Initial Force Allocations and Identify Initial Operational Risks

5-43. Planners review the forces that have been apportioned for the mission and determine if the forces available are sufficient to accomplish the mission and the specified and implied tasks. This is an initial look, recognizing that detailed force requirements cannot be determined until a concept of operations (CONOPS) is developed. But it is necessary to enable the command to identify significant force and capability shortfalls early.

5-44. In developing the campaign focus, the commander and staff focus on those elements of risk that affect accomplishment of the strategic mission. The commander must be clear as to what aspects of the campaign are critical to mission success, and where risk must be accepted or avoided. Some examples of operational risk elements might include—

- Exposure of U.S. or allied involvement inside of Country E (before a time and conditions to be determined) will result in collapse of the friendly coalition and mission failure.
- The viability of the friendly coalition will be threatened by a prolonged campaign.
- Disruption of WMD, terrorist, and human or narco-trafficking activities will provoke hostile responses unrelated to the UW campaign.
- Ethnic factors limit amount of covert resistance and U.S. activities in Country E.
- Topography, climate, and lack of infrastructure severely limit facilitating of logistic support.

Identify Initial Commander's Critical Information Requirements

5-45. Commander's critical information requirements (CCIRs) are key items of analyzed information required by the commander to make key operational decisions. They are generally tied to commander decision points (not the same as decisive points discussed above). CCIRs are dynamic—commanders add, delete, or alter CCIRs throughout the operation to help them gain clarity of the situation and in anticipation of opportunities. CCIRs should meet three criteria:

- Answering a CCIR must support a decision from the commander that staffs cannot make.
- The information or intelligence necessary to answer the CCIR must be critical to mission success.
- The commander designates the CCIR.

5-46. There are two types of CCIRs:

- *Priority intelligence requirements*: Those intelligence requirements for which a commander has an anticipated and stated priority in the task of planning and decisionmaking.
- *Friendly force information requirements*: Information that the commander and staff need about the forces available for the operation.

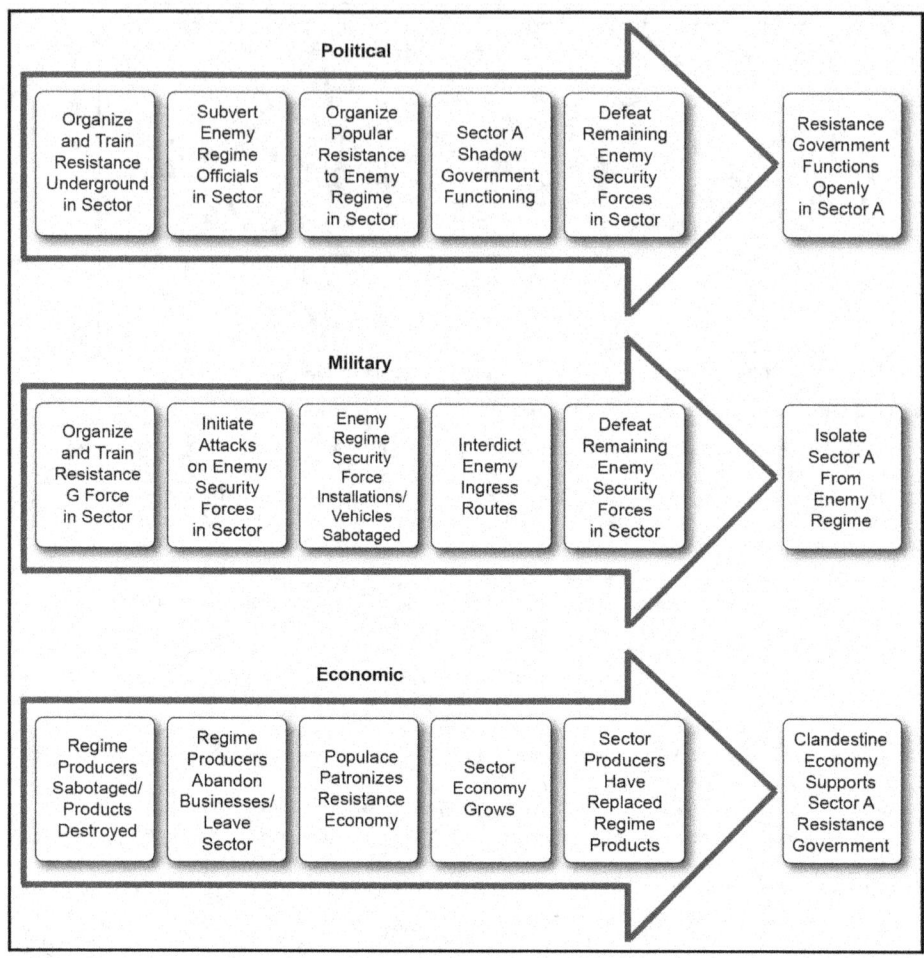

Figure 5-3. Sample lines of effort for unconventional warfare

Develop the Commander's Vision, Initial Intent, and Planning Guidance

5-47. The following paragraphs discuss this step in mission analysis.

Update Running Estimates

5-48. Staff officers should update their estimates with their analysis of the mission now that they have a better idea of what the functional requirements may be.

Develop Commander's Vision and Initial Intent

5-49. The commander now uses the understanding he has gained through his design approach, informed additionally through the mission analysis process, along with his experience, education, and wisdom, to develop an overarching vision for the campaign. This vision is the commander's personal insight of how he will employ military operations, in conjunction with interagency and multinational efforts in applying all elements of power, to achieve success. This vision, provided through commander's intent and planning guidance will facilitate military COA development, as well as proposed actions among the interagency that he believes will accomplish the national strategic end state and objectives desired.

5-50. The commander's intent is a concise narrative describing the key aspects of his understanding of the operational environment and the problem and his visualization of how the campaign must progress to achieve the desired end state. Intent focuses the staff and assists subordinates and supporting commanders in taking actions to achieve the desired end state, even when operations do not unfold as planned. Given the complexities of the operational environment at any joint level, the commander must empower subordinates to make decisions within an overall vision for success in the campaign. The commander leaves much of the detailed planning and execution of joint warfighting to his subordinate commanders and requires them to use initiative and judgment to accomplish the mission.

5-51. At the theater strategic level, commander's intent will be much broader than at the tactical level. It must provide an overall vision for the campaign that helps the staff and subordinate commanders, as well as other non-U.S. and non-military partners, to understand the intent to integrate all elements of national power and achieve unified action. The commander must envision and articulate how joint operations will dominate the adversary and support or reinforce other actions by the interagency and U.S. allies to achieve strategic success. Through his intent, the commander identifies the major unifying efforts during the campaign, the points and events where operations must dominate the enemy and control conditions in the operational environment, and where other elements of national power will play a central role. He links national strategic objectives to theater strategic objectives, and lays the foundation for the desired conditions of the military/theater end state. Essential elements of commander's intent follow:

- Purpose clearly answers the question—Why is the commander conducting this campaign? This articulation is essential not only to achieve a unity of purpose among subordinate commands, but is also crucial to provide a purpose around which military commanders may build consensus with interagency and multinational partners.
- End state specifies the desired theater strategic/military end state. Additionally, since military forces may have to support other elements of national power, the commander also explains how and when these supporting efforts will conclude at the termination of the crisis.
- Operational risk. The commander defines the portions of the campaign in which he will accept risk in slower or partial accomplishment, including a range of acceptable risk and how assuming risk in these areas may or may not impact overall outcome of the mission.

5-52. Other possible elements of commander's intent are the following:

- Objectives provide clear statements of goals of the campaign that, in combination, will lead to achievement of the military end state.
- Effects guidance provides a vision of the conditions and behaviors in the operational environment that must be in place at the successful conclusion of the campaign.
- Method provides visualization for subordinates on arrangement and synchronization of the major operations to develop future options for action. It should also explain how to support policy aims as the command becomes a supporting effort to the final achievement of the U.S. strategic ends at conflict termination.

Develop/Issue Commander's Planning Guidance

5-53. Planning guidance should enable the staff and components to understand the major themes and guiding principles for the campaign and develop detailed COAs for action. However, guidance should not be so specific as to limit the staff from investigating a full range of options for the commander. Planning guidance will provide a framework, the left and right limits, to develop options to integrate the use of military and nonmilitary power. No format for the planning guidance is prescribed.

Conduct In-Progress Review—Assumptions

5-54. At the theater level, GCCs conduct a series of in-progress reviews with the SecDef (or his designated representative) to keep the orientation of the campaign planning in line with the thinking of the national leadership. The commander considers his design, as complemented and supplemented by the staff's analysis. While the staff has been focused on the planning aspects, the commander has continued to apply a design approach to the overall situation.

5-55. Through the commander's dialogue with the national policy makers (SecDef primarily), the Chairman of the Joint Chiefs of Staff (CJCS), other supporting commands and agencies, allies, subordinate commanders, academia, think tanks, and others as he sees appropriate, he continues to enrich his understanding of the operational environment and the problem, and continues to extend and refine his visualization of the campaign. Through his experience and application of the design methodology, he is able to sense changes in the environment and refocus his understanding as appropriate. He transmits this increased understanding and visualization to his staff and subordinates as often as he believes there is need for updated focus. The GCC brings his most recent understanding and visualization to the SecDef in-progress review. The SecDef will approve the GCC's mission statement and provide further guidance as required to guide continued design and planning.

STEP #3: COURSE OF ACTION DEVELOPMENT

5-56. One way staffs help commanders refine their visualization is to develop alternative COAs, execute the commander's envisioned operational approach, and achieve the objectives. For each COA, the staff must enable the commander to envision the employment of friendly forces and assets as a whole, taking into account externally imposed limitations, the factual situation in the AO, and the conclusions from mission analysis. Equally important, the commander must envision how military forces will work in conjunction with the other elements of national power to achieve military and strategic ends.

5-57. For each COA, staffs develop a narrative and sketch that provides the following:
- Objectives.
- Key tasks.
- Major capabilities required.
- Task organization.
- Main and supporting efforts.
- Sustainment concept.
- Deployment concept.
- Informational themes.
- Identification of reserve (if appropriate).
- Identification of major interagency supporting tasks.

Determine Opposing Courses of Action

5-58. Before developing possible COAs, the staff must gain an appreciation of what other actors may do to shape the future operational environment to their desired end state. Staffs must consider not only enemy and adversary actions, but also neutral and friendly actions that may impede achievement of the friendly desired end state. It is important to determine how the enemy will attempt to accomplish their strategic goals by identifying likely objectives and desired end states, potential strategic and military capabilities, and estimate how the opposition leadership may apply his elements of power in the future—the opposing COAs. This requires identifying all known factors affecting the opposition's actions, and requires the commander and his staff to think as the opponent thinks.

5-59. The staff will identify for the commander both the most-dangerous opposing COA, as well as the most-likely opposing COA, based upon the situation anticipated and/or at hand. Usually, commanders consider the most-likely opposing COA as their baseline for friendly action unless the consequences of not focusing on the most-dangerous opposing COA preclude doing otherwise. Regardless of which opposing COA supports the baseline planning effort, staffs must develop branches for the others, as time permits. This analysis will not only influence the JPG's development of COAs, but also will form the basis to focus and develop priority intelligence requirements.

Refine the Commander's Operational Approach to Develop an Initial Framework

5-60. The JPG will analyze the commander's guidance to develop a more detailed framework of nested objectives and effects for accomplishment during the campaign to achieve the military end state. In refining the commander's design, the staff analyzes how the broad, overarching guidance for the campaign will break down into more detailed and achievable blocks as the campaign unfolds. This analysis of nested objectives and effects provides a framework for the logical development of tasks by components and functions that will achieve the desired conditions in the operational environment. Effects serve as a good translation of objectives into tasks by providing a statement of the conditions desired (and undesired) in order to achieve objectives.

Develop Courses of Action

5-61. The JPG develops and analyzes a range of potential military and nonmilitary actions, and assesses how well each of these accomplish the desired effects on the operational environment, given the time and resources available. Some guidelines for development of COAs follow:
- Develop a variety of diverse and distinguishable options to accomplish the commander's vision so as to provide a broad range of options for his consideration and selection.
- Employ the elements of operational design as a framework for development and analysis of how to sequence, synchronize, and integrate joint operations.
- Sequence and focus joint functions to accomplish the tasks required to dominate and control decisive points.
- Ensure clearly articulated objectives and effects as the guideline for actions during each phase or period of the campaign.
- Provide only valid options that are suitable, feasible, and acceptable, based upon time, forces and capabilities, and resources available, and that fall within acceptable levels of operational risk.

5-62. In developing, analyzing, and proposing COAs for the campaign, the commander and JPG analyze not only military actions, but all instruments of national power to determine how best they can bear against the adversary's instruments of power (political, diplomatic, informational, military, and economic) to achieve military and strategic ends. Potential COAs may vary the use of forces (Army, Navy, Marine, joint, and so on), differing timing and sequencing of operations, and the use of other instruments of power (information, economic, diplomatic) in combination with military functions. One possible method of visualizing sequencing during COA development is to organize tasks and lines of operation into prehostilities, hostilities, and post-hostilities periods, vice more detailed phases.

Develop an Initial Concept Graphic and Narrative

5-63. The following sequential steps can help the planning staff in building the sketch and narrative:
- Determine forces available/apportioned.
- Post decisive points. Review the operational centers of gravity as the point of focus for operations and post the major, physical, and logical decisive points that will be relevant to the COA.
- Array forces at military end state. Position forces geographically where they are needed in the theater at the end of the campaign and determine what those forces will do.

Chapter 5

- Identify initial entry points. Based on initial guidance and knowledge of theater access and facilities, display where the forces can enter the theater from air and sea deployments, and show the initial bases/staging areas available to support this deployment. Also portray the initial lines of communication that will connect initial forces back to in-theater (intermediate staging bases) and strategic (continental United States or forward-deployed) bases of operations.
- Array forces at prehostilities. Formulate the initial concept for a basing plan and joint reception, staging, onward movement, and integration (JRSOI).
- Maneuver the forces forward to end state. Looking at the sketch with the end state and objectives or effects by period or phase in mind, determine the best way to get the forces into theater and to their ultimate locations at the end of the campaign from bases in friendly territory. This will help formulate the desired basing plan for the beginning, middle, and end of the campaign.

Determine the Tasks Required in Each Period

5-64. Planners analyze how military and nonmilitary actions will accomplish the required changes in the operational environment. It is important to also identify tasks for interagency partners (DOS, Department of Treasury, and so on), coalition and international organizations (United Nations, regional organizations), and other nongovernmental partners. The staff focuses on the effects desired or to avoid, and considers how to employ air, land, maritime, and special operations forces, in conjunction with other forces and elements of power for intelligence, protection, projection, theater opening, sustainment, and IO. Considerations for tasks include—

- Conducting initial entry into theater: basing, access, and overflight.
- Deploying and receiving of the force.
- Protecting forces and host-nation points of entry.
- Building and maintaining a coalition force.
- Achieving command and control with joint, host-nation, and coalition forces.
- Achieving the desired effects.
- Preventing undesired effects/events, such as a humanitarian crisis, loss of local support, and so on.
- Sustaining the joint force and additional support required to enable and maintain host-nation and coalition participation.
- Determining post-hostilities conditions, and how the joint force will maintain military gains and transform them into long-term strategic success.

Identify Main and Supporting Efforts

5-65. At this point in initial concept development, there is no need to specify who is the main effort, but identifying what is the main effort is important.

Test the Courses of Action for Validity

5-66. Before going further in developing the COAs, planners determine if the COA meets the validity criteria:

- *Adequate*: Accomplishes the mission within guidance; meets end state, objectives, and effects for campaign.
- *Feasible*: Can execute within time, space, and resources available.
- *Acceptable*: Achieves ends balanced with costs/risk.
- *Distinguishable*: The COA is different from others; for example, vary key elements such as force structure, defeat mechanism, operational design elements, and so on.
- *Complete*: Incorporates all key elements.

Determine Initial Mission Command Relationships

5-67. Planners will have to identify the basics of how components will be organized, any JTF requirements, and how the joint force will control or coordinate its efforts with the host nation, multinational forces, and interagency elements as necessary.

Update Running Estimates

5-68. Because of the unique talents of each joint staff directorate, involvement by all is vital to the process. Each running estimate takes on a different focus that identifies certain assumptions, detailed aspects of the COAs, and potential deficiencies that are simply not known at any other level, but nevertheless require detailed consideration.

Provide Initial Course of Action Brief to the Commander

5-69. The goal is to provide the commander an azimuth check before proceeding into COA analysis, and to gain insights on whether the work thus far meets guidance. At the end of this briefing, the staff must know which COAs should move forward for further analysis and development, with additional guidance on modifications, improvements, and/or risk.

STEP #4: COURSE OF ACTION ANALYSIS

5-70. Once the staff has completed COA development and an azimuth check with the commander, the JPG analyzes each COA in detail. The objective of this step is to analyze each COA critically, independently, and according to the commander's guidance, in an effort to determine the advantages and disadvantages associated with each COA.

5-71. Wargaming is a simulation of a military operation involving two or more opposing forces, using rules, data, and procedures designed to depict an actual or assumed real-life situation. It is a conscious effort to visualize the flow of a plan, within an operational environment, using joint forces, while integrating the other elements of power as appropriate, and a realistic, thinking, and adaptive adversary. Wargaming assists joint-force planners to identify the strengths and weaknesses, associated risks, and asset shortfalls for each friendly COA. While joint doctrine refers to visualizing the flow of a military operation as the key element in wargaming, the commander and staff must consider the application of all elements of national power (diplomatic, informational, military, and economic).

Develop Governing Factors

5-72. Determining the initial evaluation criteria is a critical requirement that begins before COA analysis. Some examples of potential governing factors are—
- Risk (during operations and after; strategic and operational).
- Flexibility.
- Time.
- Sustainment and support.
- Surprise.
- Force protection.
- Casualties.
- Use of flexible deterrent options.
- Financial costs.
- Impact on coalition interests.

War Game Each Course of Action Independently

5-73. Wargaming provides a useful means for the commander and staff to analyze and test each friendly COA against a selected opposing COA in an action-reaction-counteraction methodology. The COAs must

Chapter 5

be evaluated through the other actors' eyes, given their political and cultural perspectives and biases, in order to determine if the proposed actions will change the intended behaviors in the manner that friendly planners believe—a key aspect to achieve desired, rather than undesired, effects. In addition to actions by adversaries, actions by neutral or even friendly actors may be counterproductive to the commander's objectives, needing to be considered effectively as opposing actions. While the main effort is on applying the use of military power, consider all elements of power available. While the commander may not be able to control the diplomatic, informational, and economic actions, he can coordinate these elements with other actors who may be able to influence the application of the other elements of power.

Decide What Type of War Game to Use

5-74. Wargaming has manual and computer-assisted components. Manual wargaming makes up the bulk of activity when staffs war game. Automation normally resolves questions regarding outcomes during specific moments in the fight, to determine the gross requirements for each class of supply, and to conduct initial strategic, transportation feasibility. Automation can never supplant the combined experience of the people conducting the war game. When time and automated resources are lacking, manual-only wargaming will suffice. There are several approaches to visualize the flow of the war game. The preference of the commander, the scope and nature of the plan, and the level of sophistication of the JPG determine which approach to use. Visualization methods include—

- Major periods construct with prehostilities, hostilities, and post-hostilities.
- Phasing model articulated in joint doctrine or another phasing model developed for the campaign.
- Sequencing of critical events, decisive points, or major tasks.

Conduct the War Game

5-75. Planners prioritize the opposing COAs for analysis against the friendly COAs and begin. A simple manual war game method employs an action-reaction-counteraction format between blue, red, and green (neutral actor) teams. A possible framework to guide the flow is to use the lines of operation or lines of effort to sequentially work through the campaign. As the JPG conducts the war game, they interpret the results of analysis to ensure each COA remains valid. If a COA is inadequate, infeasible, or unacceptable, they must discard or modify the COA. Throughout the analysis and wargaming process, the JPG must remain focused on the following areas:

- Objectives: strategic, theater strategic, and operational.
- Balance between creativity and the realities of the operational environment.
- The elements of operational design.
- Joint functions.

Record the War Game

5-76. Proceedings of the war game can be recorded in a variety of means:

- Narrative describing the action, probable reaction, counteraction, assets, and time used.
- Sketch note which uses a narrative but adds operational sketches to paint a clear picture.
- Matrix worksheet with action, reaction, counteraction, assets, and time columns, and major event rows.
- Synchronization matrix organized by time or major events as columns, with functional and other major activity areas as rows. If used as a recording tool, this would form the beginning of the synchronization matrix that will provide the commander and staff a visualization tool for the campaign. It can be refined throughout planning, and should be updated throughout the campaign. The synchronization matrix helps staff officers build the detailed functional plans that support the campaign plan.

5-77. Whichever method of recording the war game is used, it is important to capture the decision points, CCIRs, COA adjustments, potential branches and sequels, and potential undesired effects.

Analyze the Course of Action

5-78. The analysis of the COA as a result of the war game should include the following areas: propensity to achieve the desired operational environment; advantages and disadvantages; critical events, decision points, and CCIRs; potential branches and sequels; risks of undesirable effects; and strategic challenges that must be resolved. Once analysis is complete, adjust the COA to mitigate risk and enable it to better achieve objectives and update the running estimate.

STEP #5: COURSE OF ACTION COMPARISON

5-79. After rigorous independent analysis of each COA, the JPG compares the COAs with one another using a common set of criteria. Using evaluation criteria (governing factors) derived mostly from his intent and guidance, the staff analyzes and evaluates the COAs against the commander's standards—not against one another—in order to identify the one that best meets the commander's needs.

Compare Courses of Action Using Governing Factors

5-80. The COAs are compared using the evaluation criteria that was established prior to the wargaming, and probably augmented as a result of wargaming. The staff should remain as objective as possible when comparing the COAs and be careful of manipulating criteria to promote a favorite COA. Weighting evaluation criteria is a frequent and often helpful technique to identify the most-critical criteria. Weighting, like evaluation criteria selection, should come prior to formal COA comparison to avoid assigned weight manipulation.

Balance Ends, Ways, Means, and Risks of Each Course of Action Against the Others

5-81. COAs should be compared to one another using the governing factors, but also using a holistic look that considers risk to the force, force posture and readiness at the end of the campaign or operation, degree of initiative left to subordinates to react to unforeseen circumstances, and flexibility to react to unexpected threats and opportunities.

Select the "Best" Staff-Recommended Course of Action

5-82. After the comparison analysis, the staff must select the COA that they will recommend to the commander. This selection must consider not only the JPG analysis, but each staff section's functional analysis of the COAs. COA comparison is a subjective process and should not become a mathematical exercise, though using +, -, 0, or 1, 2, 3 as expressions of relative value may be appropriate. The key element in this process is the ability to articulate to the commander why one COA is preferred over another in terms of how well the COA meets the evaluation criteria requirements.

STEP #6: COURSE OF ACTION APPROVAL

5-83. The JPG briefs the results of the wargaming analysis and the COA comparison analysis to the commander to obtain a commander's decision on which COA to develop into the CONOPS of the campaign. This enables the commander to refine his visualization of the campaign and provide further guidance to the staff on how to proceed with CONOPS development. It also prepares the commander for another in-progress review with the SecDef (or his next-higher command).

Recommend Course of Action to the Commander

5-84. The staff presents the COA analysis and the recommended COA. The forum to present the results of COA comparison is the commander's decision brief (Figure 5-4, page 5-22). Typically, this briefing provides the commander with an update of the current situation, an overview of the COAs considered, and a discussion of the results of COA comparison. This decision brief will also include an update on the understanding of the operational environment that was enabled through the wargaming.

1. State purpose of the briefing.
2. Update the understanding of the operational environment.
3. Update the friendly situation (military forces and other relevant elements of power).
4. Provide mission statement.
5. Update assumptions/facts.
6. Present each course of action:
   - Narrative and/or sketch.
   - Advantages/disadvantages/risks.
   - Needed branches, changes, resources.
7. Compare courses of action:
   - Relative advantages/disadvantages.
   - Comparison using evaluation criteria.
8. Recommend course of action.

Expected outcome:
- Commander's approval of a course of action or approval of a modified course of action.
- Update on the commander's visualization of the campaign.

Figure 5-4. Course of action decision brief

5-85. During the brief, it is important that dissenting views be heard so the commander can hear all aspects of the analysis. Staff officers should be encouraged to expound on issues in their functional areas if needed. Subordinate commands should be present or linked via video-teleconference. Other partners should also be invited to the brief, to include OGAs and key multinational partners, to the extent possible or appropriate. Staff officers from those organizations are probably part of the JPG, so there should be no surprises.

Commander Selects a Course of Action or Approves a Modified Course of Action

5-86. The commander will evaluate all analyses of the JPG, applying his own understanding of the operational environment and the mission and his visualization of the campaign, to evaluate critically how each COA would accomplish the mission. The commander may select a single COA as presented, or may incorporate the best portions of several COAs to form a new one.

Receive Commander's Guidance for Concept Development

5-87. As part of the COA decision brief, or following it, the commander will likely provide additional guidance that will enable the development of the CONOPS and the strategic plan. Once the COA is approved, it now becomes the commander's CONOPS. Confirm any updated commander's intent.

Update Running Estimates and the Commander's Visualization

Once the commander makes a decision on a COA, staff officers refine their estimates of the campaign's supportability from their functional viewpoint. The staff also prepares the commander's visualization, which provides a concise statement of how the commander intends to accomplish the mission and provides the necessary focus for continued campaign planning and developing an operation plan (OPLAN)/operation order (OPORD) (Figure 5-5, page 5-23). At the combatant-command level, this visualization goes to the CJCS for SecDef review and is the basis for a decision on which COA to refine and potentially execute.

> Operational description:
> - Purpose of the operation.
> - References.
> - Description of military operations.
>
> Narrative:
> - Mission.
> - Situation and courses of action.
> - Analysis of opposing courses of action.
> - Comparison of friendly courses of action.
> - Recommendation (or decision).
>
> Remarks. Cite plan identification number of the file where detailed requirements have been loaded into the Joint Operation Planning and Execution System (or Adaptive Planning and Execution System).

Figure 5-5. Commander's visualization

Conduct In-Progress Review—Concept

5-88. During in-progress review—concept (IPR-C), the SecDef (or his representative) will consider the GCC's analysis and approve (or modify) the CONOPS for further development. Based upon the SecDef's decision and further strategic guidance, the GCC will refine his CONOPS and reissue his intent and planning guidance to drive development of the plan during the next step of the process.

STEP #7: PLAN OR ORDER DEVELOPMENT

5-89. After the commander has approved a COA, and provided additional guidance to the staff for development of the CONOPS and the full plan (with updates as required after the IPR-C for combatant commands), the staff develops the CONOPS into an operations plan or operations order. The CONOPS must be developed to provide the detail required for the staff to build the base plan and prepare supporting annexes, and supporting and subordinate organizations to build supporting functional plans.

Review Planning Guidance and any Updated Commander's Intent

5-90. The staff should review the commander's guidance as updated throughout the planning process and as modified as a result of the IPR-C and associated discussions by the commander. The commander should republish his intent, with any changes to it that may result from his increase in understanding of the operational environment and the problem and his vision for the campaign.

Phase the Concept—UW Considerations of the Phasing Model

5-91. Although the commander will determine the number and actual phases used during a campaign or operation, use of the phases described Figure 5-6, page 5-24, provides a flexible model to arrange combat and stability operations. Depending on the purpose and operational design, a UW campaign could directly follow this six-phase model. However, planners should acknowledge that the standard Phase V (Enable Civil Authority) describes a return to more normal patterns of civil governance and support to already existing legitimate indigenous institutions. Therefore, at some point within Phase IV (Stabilize), the campaign transitions from UW to foreign internal defense. The six-phase model is not intended to be a universally prescriptive template for all conceivable joint operations and may be tailored to the character and duration of the operation to which it applies.

Chapter 5

Phase 0—Shape. Joint and multinational operations— inclusive of normal and routine military activities—and various interagency activities are performed to dissuade or deter potential adversaries and to assure or solidify relationships with friends and allies.

Phase I—Deter. The intent of this phase is to deter undesirable adversary action by demonstrating the capabilities and resolve of the joint force. It includes activities to prepare forces and set conditions for deployment and employment of forces in the event that deterrence is not successful.

Phase II—Seize Initiative. Joint force commanders seek to seize the initiative through the application of appropriate joint force capabilities.

Phase III—Dominate. The dominate phase focuses on breaking the enemy's will for organized resistance or, in noncombat situations, control of the operational environment.

Phase IV—Stabilize. The stabilize phase is required when there is no fully functional, legitimate civil governing authority present. The joint force may be required to perform limited local governance, integrating the efforts of other supporting/contributing multinational organization, international organization, nongovernmental organization, or U.S. Government agency participants until legitimate local entities are functioning.

Phase V—Enable Civil Authority. This phase is predominantly characterized by joint force support to legitimate civil governance in theater. The goal is for the joint force to enable the viability of the civil authority and its provision of essential services to the largest number of people in the region.

Figure 5-6. Joint Operation Planning and Execution System phasing model

5-92. Within the context of these phases established by a higher-level JFC, subordinate JFCs and component commanders may establish additional phases that fit their CONOPS. For example, the joint force land component commander (JFLCC) or a subordinate JTF might have the following four phases inside the GCC's seize initiative phase: deploy, forcible entry, defense, and offense. The JFLCC could use the offense phase as a transition to the GCC's dominate phase. This "phases within higher phases" relationship is typical of and particularly relevant to UW planning. The entire seven phases of UW could fit within select parts of a higher JFC's phased plan. For example, the UW employment phase (the 6th of the 7-phase UW model) can be coordinated to provide critical synchronized support to conventional forces' main effort in Phase II (Seize Initiative) and Phase III (Dominate). Alternately, UW could be particularly useful in the early phases of a joint campaign to shape and deter events in Phases 0 and I. Successful UW conducted in these early phases may preclude the need to progress to the later phases of more costly and intrusive conventional-force involvement. This latter case is also true when there is no friendly intent to dominate an opponent—as in the relatively rare case of complete overthrow—but only to coerce or disrupt an opponent for relatively limited objectives.

5-93. It must be emphasized that each step in the six-phase JOPES model and the seven-phase model of UW do not correspond directly or automatically. For example, JOPES Phase 0 (Shape) is not synonymous with UW Phase 1 (Preparation), nor is JOPES Phase III (Dominate) synonymous with UW Phase 6 (Employment). The relationship—if any—between these two templates will vary with the design of each specific campaign and the explicit authority from the President or SecDef to conduct a GCC's OPLAN for a UW campaign. This means that many activities which are commonly associated with UW, such as PE, may be conducted routinely and for long periods under other plans during the JOPES Phase 0 (Shape) without being UW.

5-94. Each phase of the campaign's CONOPS should describe the following elements:
- Intent and CONOPS for the phase. The commander's intent for the phase must be clear. This section describes the purpose, end state, and a description of the operational risk to the campaign during this phase. The description of the CONOPS may be a narrative of the various lines of operation and lines of effort as they are executed during this particular phase.
- Objectives and effects (desired and undesired). This section describes the objectives for the phase, and the major effects that must be achieved to realize the objectives. It describes how the

force's objectives are related to those of the next-higher organization and to other organizations (especially if the military is a supporting effort).
- Tasks to subordinate commands and supporting commands and agencies. The commander assigns tasks to subordinate commanders, along with the capabilities and support necessary to achieve them. Area tasks and responsibilities focus on that specific area to control or conduct operations. Functional tasks and responsibilities focus on the performance of continuing efforts that involve the forces of two or more Military Departments operating in the same domain, air, land, sea, or space, or where there is a need to accomplish a distinct aspect of the assigned mission. Identification of requests to organizations outside of DOD for support is included.
- Mission command organization and geometry of the AO. Any changes to the command and control structure or to the geometry of the AOR are noted.
- Assessment methodology. This section identifies the basic methodology for assessing accomplishment of objectives. Assessments are also included to help gauge if the objectives actually support achievement of the end state.
- Risk mitigation. This section identifies the areas of risk concern to the commander and outline how the risk may be mitigated.
- CCIRs and associated decision points.
- Transition to the next phase. This section describes how the joint force will move to the next phase. It describes the end state conditions for the phase, which should tie directly to the initiation conditions for the next phase. It also includes a description of transition of control from the joint force to other parties for aspects of the overall campaign.

Develop Supporting Functional Concepts

5-95. Supporting concepts are built to ensure supportability and coordination among all of the functions. Some of the key functional concepts are for IO/inform and influence activities (IIA) and MISO, CMO, logistics support, force projection, IO/IIA, force protection, and logistics and sustainment. At the combatant command level, the CJCS issues a planning order or alert order to direct preparation of supporting plans after receipt and approval of the commander's COA as transmitted in the commander's visualization. The command informally coordinates with organizations outside of DOD to build mutually supporting concepts. The command will integrate these concepts into the campaign plan.

Information Operations/Military Information Support Operations Considerations in Unconventional Warfare Planning

5-96. IO/IIA and MISO plans outline JFC coordination and integration with interagency activities in every phase of a UW campaign. The specific requirements of such coordination and integration will be tailored to the requirements of each phase. For example, the IO/IIA and MISO concept of support for Phase I, Preparation, will focus on PE for follow-on UW activities. The IO/IIA and MISO concept of support for Phase II, Initial Contact, will focus on the activities, capabilities, and authorities necessary to enable initial contact with resistance groups, and so on throughout all seven phases.

5-97. Each phase of the IO/IIA and MISO concept of support will also identify the IO/IIA objectives intended for each phase of the larger campaign plan. For example, IO/IIA objectives in Phase III, Infiltration, may include a continuation of ongoing diplomatic efforts to support friendly activities while seeking to deny political support to the target of the UW campaign. It will likely include the full range of overt, covert, and clandestine IO/IIA and MISO activities that support the infiltration of friendly forces into the JSOA.

5-98. Each phase of the IO/IIA and MISO concept of support will specify the different target audiences that the coordinated IO/IIA and MISO plan is intended to affect. For example, typical target audiences of a UW campaign will include governments and public audiences outside of the JSOA that have a potential to assist or harm friendly UW activities. Inside of the JSOA, target audiences may include the resistance or insurgent elements receiving U.S. assistance; significant demographic groups —as determined by careful population analysis—that could potentially support or harm the resistance or insurgency and the target of the UW campaign; the regime (or occupying power), its armed forces, and security apparatus; the bureaucracy; and other formal or informal actors who provide regime/occupying power support.

5-99. For each target audience, IO/IIA and MISO tasks to be completed and/or IO/IIA effects to be achieved will be specified for each target audience by phase of support. Finally, any authorities required to achieve each task to be completed and/or effect desired will be identified by phase of support. An example of a blank, generic IO/IIA and MISO support plan format is shown in Figure 5-7. FM 3-53 provides further detail on authorities.

Note: Not all required authorities listed may be granted at the time of planning; nevertheless, identification and explanation of such authorities can function as critical information for decisionmakers as the plan develops. This format is only a minimal guide and planning aid; the UW campaign design should not be constrained by the conceptual limits of the format.

Information Operations/Inform and Influence Activities Concept of Support: "Coordinate/integrate with interagency activities, capabilities, and authorities to ..." (provide whatever information operations/inform and influence activities and Mililitary Information Support Operations support is pertinent to the requirements of the larger unconventional warfare campaign plan during this phase).		
Information Operations/Inform and Influence Activities Objectives: (List the few key information operations/inform and influence activities and Military Information Support Operations objectives sought in this phase of the larger unconventional warfare campaign plan).		
Target Audiences (Examples:)	**Tasks - Effects (Examples:)**	**Authorities (Examples:)**
A list of governments and populations of countries in vicinity of or influencing joint special operations areas; international organizations, nongovernmental organizations, and so on.	Diplomatic activities in coalition building; strategic communications and public diplomacy; defense support to public diplomacy; third-party initiatives; a list of classified activities; and so on.	A list of pertinent U.S. and international laws, including relevant sections of U.S. Code; classified authorities (see U.S. Special Operations Command Directive 525-89).
A list of resistance/insurgent groups and population subgroups, joint special operations area, internal target audiences (based on demographic analysis).	A list of specific unclassified and classified activities tailored to influence appropriate target audiences.	A list of pertinent U.S. and international laws, including relevant sections of U.S. Code; classified authorities (see U.S. Special Operations Command Directive 525-89).
A list including the targeted regime or governing authority, government institutions, military and security forces and support structures, and so on.	A list of specific unclassified and classified activities tailored to influence appropriate target audiences.	A list of pertinent U.S. and international laws, including relevant sections of U.S. Code; classified authorities (see U.S. Special Operations Command Directive 525-89).

Figure 5-7. Information operations/Military Information Support Operations concept of support plan format

5-100. The final section of the IO/IIA and MISO concept of support plan consists of a list of SOF IO/IIA strengths and capability/authority gaps relating to the execution of the campaign plan. Such lists provide decisionmakers a snapshot of continuing or pending shortfalls in needed requirements for plan execution.

Civil Affairs Operations Considerations in Unconventional Warfare Planning

5-101. CAO is a critical component of UW. The process of obtaining and analyzing civil information on infrastructure and conditions, identifying civil vulnerabilities and centers of gravity, gaining and maintaining access to key areas and populations, identifying and establishing relationships with key civic and skilled personnel, fusing civil information with intelligence, and influencing key areas and relevant individuals, groups, and populations begins in Phase I and continues through Phase VII.

5-102. CA forces focus on identifying civil vulnerabilities to be either exploited or addressed by the shadow government and its guerrilla movement. Further mentorship concerning provision of goods and services to deprived populations by guerrilla forces is a capability of CA forces. Ultimately, UW is won not by military actions conducted by guerrilla forces, but by popular opinion of the shadow government versus the current regime; thorough, comprehensive CAO will significantly increase the shadow government's standing in the war of opinion.

5-103. UW planners need to outline CAO and CMO within each phase of the operation. When feasible, throughout all phases of the campaign, DOD elements should work with indigenous and interagency counterparts (for example, USAID) for governance, development, and consequence management. An indigenous lead allows for a sustainable and legitimate CMO effort and reduces the need for a high-profile U.S.-led effort. A generic by-phase concept outline of expected CMO support is shown in Figure 5-8.

Civil-Military Operations Concept of Support to Unconventional Warfare Campaign
Phase I—Preparation. Conduct an area assessment (to include an assessment of unconventional warfare potential) from a Civil Affairs perspective; concentrate particularly on a political, military, economic, social, information, and infrastructure systems analysis approach to economic and infrastructure factors, and other relevant information; conduct civil-military engagements in order to develop and exploit relationships with resistance expatriates and refugees.
Phase II—Initial Contact. Establish contact with the resistance; conduct civil reconnaissance and humanitarian assistance/ civil-military operations in refugee camps when applicable in order to gain access and identify persons of influence/shadow government. Engagement plans should be updated and detailed as new information is acquired.
Phase III—Infiltration. Train and assist resistance to conduct civil-military operations and advise and assist shadow government in order to expand popular support for the resistance.
Phase IV—Organization. Conduct civil-military operations and training of the shadow government and refugee populations in order to build capacity of shadow government.
Phase V—Buildup, and **Phase VI—Employment.** Increase domestic support for the resistance groups by providing civil-military operations training and humanitarian assistance to diaspora and refugee populations; provide clandestine civil-military operations and humanitarian assistance support to resistance elements as they increase clandestine reach into new populations to increase legitimacy of the resistance; monitor and assess effects and conduct consequence management in order to establish credibility of the shadow government.
Phase VII—Transition. Promote the new government's legitimacy; transition from a unconventional warfare task force to a security force assistance organization; increase support to civil administration.

Figure 5-8. Civil-military operations concept of support plan format

Other Supporting Concepts

5-104. IO/MISO and CMO are organic, traditional, and mandatory ARSOF contributors to any UW effort. The contribution of other supporting actors and agencies will vary, however, depending on the given situation. Any JIIM partner can and should have a written concept of support to the UW campaign plan:

- *Joint.* By definition, UW is conducted among populations on land, and the U.S. Army SF are the unique unit specifically designated to conduct UW under the proponency of the USASOC. That does not mean, however, that other components do not conduct or support UW. Navy, Marine, and Air Force component participation in UW campaigns will be determined by the specific conditions pertaining to the situation. Naval special operations command, Marine Corps special operations command, and Air Force Special Operations Command all have core or derived capabilities to conduct or support a UW effort. When components play a major role in executing UW, the articulation will be a major maneuver portion of the CONOPS in addition to a supporting concept product.
- *Interagency.* All U.S. official international interactions, including support to a UW campaign, will involve the DOS participation. All U.S. UW campaigns will involve support from U.S. civilian intelligence agencies—most crucially in the Phase 0 preparations in the operational area prior to U.S. military involvement. Most other USG agencies (for example, Treasury, Justice, or Commerce) can also potentially play a role in support of a UW campaign.

Chapter 5

- *Intergovernmental.* Most intergovernmental support will be coordinated and facilitated by and through the DOS. Foreign government support is essential to virtually any UW effort and such support should be articulated in a supporting concept. Although some foreign support may be too sensitive for inclusion in planning documents, the support concept will be outlined and made available to national leadership and the GCC.
- *Multinational.* Not all foreign support is derived from foreign governments. The global scope and often unparalleled reach of nongovernmental organizations make them a potentially valuable contributor to UW campaigns. Sometimes, nongovernmental organization involvement can also be too sensitive for inclusion in planning documents but will likewise be outlined and made available to national leadership and the GCC.

5-105. Synchronization of the plan takes place once all of the supporting concepts are built. Synchronization is the art of arranging all activities (military and otherwise) in the right sequence and place, with the right purpose, to produce maximum effect at the decisive points. Synchronization will continue after development of the plan, through briefbacks, rehearsals, and execution.

Expand the Concept of Operations into a Base Plan With Annexes

5-106. As the CONOPS develops into a fully detailed plan, a number of activities coincide in a parallel, collaborative, and iterative fashion rather than in a sequential and time-consuming manner. Planners ensure that the effects of all component plans are synchronized to meet the campaign end state.

Force Planning

5-107. Force planning begins early during concept development but must be refined and finalized during detailed planning. The commander determines force requirements, develops a letter of instruction for time phasing and force planning, and designs force modules to align and time-phase the forces IAW the concept under development. Major forces and elements initially come from those apportioned or allocated for planning by operational phase, mission, and mission priority. Service components then collaboratively make tentative assessments of the specific combat and supporting capabilities required. The commands should not be constrained by the apportioned forces, but must be able to provide clear rationale for capabilities required that are not apportioned. The commander typically describes required force requirements in the form of broad capability descriptions or unit type codes, depending on the circumstances.

Support Planning

The purpose of support planning is to determine the sequence of the personnel, logistics, and other support required to provide distribution, maintenance, civil engineering, medical, and other sustainment IAW the CONOPS. Support planning takes place in parallel with other planning. Support planning is primarily the responsibility of the Service component commanders. Service component commanders identify and update support requirements in coordination with the Services, the Defense Logistics Agency, and U.S. Transportation Command. They initiate the procurement of critical and low-density inventory items, determine host-nation support availability, develop plans for total asset visibility, and establish phased delivery plans for sustainment in line with the phases and priorities of the concept. They develop battle damage repair programs, reparable retrograde plans, container management plans, force and line-of-communications protection plans, supporting phased transportation and support plans aligned to the strategic concept, and report movement support requirements. Service component commanders continue to refine their sustainment and transportation requirements as the force providers identify and source force requirements. The requirements and transportation planning must be integrated and coordinated by the GCC to ensure synchronization with the CONOPS, to reduce redundancies and manage risk, and to integrate transportation requirements with the force flow.

Deployment Planning

5-108. The anticipated operational environment dictates the type of entry operations, deployment concept, mobility options, predeployment training, and force integration requirements. The GCC is responsible for developing the deployment concept and identifying predeployment requirements. The combatant command is also responsible for movement planning, manifested through the time-phased force deployment data file, assisted by the force providers and U.S. Transportation Command. In particular, U.S. Transportation Command assists greatly with current analysis and assessment of movement command structures and systems; available organic, strategic, and theater lift assets; transportation infrastructure; and competing demands and restrictions. All recognize that operational requirements may change, resulting in changes to the movement plan. Planners must understand and anticipate the physical limitations of movement assets and infrastructure, and the impact of change, as any change will have an effect on the rest of the time-phased force and deployment data. Finally, the supported command is responsible for JRSOI planning. JRSOI planning ensures an integrated joint force arrives and becomes operational in the AO as required.

Shortfall Identification

5-109. The supported commander continuously identifies limiting factors and capabilities shortfalls and associated risks as plan development progresses. If internal resolution is not possible, the supported commander reports these limiting factors and assessment of the associated risk to the CJCS. The CJCS and the Service chiefs coordinate resolution, with one option being acceptance of risk.

Feasibility Analysis

5-110. The focus in this activity is to ensure assigned mission accomplishment, using available resources within the plan's contemplated time frame. Measures to enhance feasibility include adjusting the CONOPS, ensuring sufficiency of resources and capabilities, and maintaining options and reserves.

Complete Coordination of the Plan

5-111. Supported commands review all of the supporting plans once they are prepared to ensure that the plan is fully coordinated. Planning for multinational operations is coordinated through various means. Individual treaty or alliance procedures set the stage for collective security goals, strategies, and combined operation plans IAW U.S. doctrine and procedures. Thus, much information and guidance for joint operations are conceptually applicable to alliance and coalition planning.

5-112. Coordination of the plan with interagency partners is conducted both informally and formally. GCCs and JFCs should encourage and solicit maximum participation of appropriate interagency planners in the design of campaigns and operations. Their participation throughout planning is extremely beneficial to expand the perspectives and expertise provided in design and in achieving unity of purpose and then unity of effort in the campaign or operation.

Brief the Plan for Approval

5-113. Once completely coordinated, the plan should be briefed through to the commander for his validation, as well as to prepare him to brief the plan to the national leadership.

Complete In-Progress Review—Final

5-114. Once the plan is completed, the GCC submits it with the associated time-phased force and deployment data file to the Joint Staff for review. The joint planning and execution community reviews the plan for—

- *Adequacy.* Does the plan satisfy the mission and comply with guidance provided?
- *Feasibility.* Are the required resources available in the time frames anticipated?
- *Acceptability.* Are the anticipated operations proportional and worth the anticipated costs? Is it politically supportable?
- *Completeness.* Does the plan include all required parts?
- *Compliance.* Does the plan comply with joint doctrine?

5-115. In conjunction with the GCC's plan brief at the in-progress review—final, the CJCS and the Under Secretary of Defense for Policy will also offer their military advice. This advice includes identification of national military strategic issues arising from, or resolved during, plan review, such as key strategic risks and national-level decision points.

5-116. The result of the in-progress review—final is SecDef approval of the base plan and required annexes, the resolution of any remaining key issues, and approval to proceed with plan assessment, as applicable, with any amplifying guidance or direction.

Issue the Operation Plan or Operation Order

5-117. If the plan is issued as an operation order, it will be immediately transitioned to execution. A transition brief provides an overview of the current and desired operational environment, mission, commander's intent, and vision to provide information and direction to those who will execute the campaign. The brief should include items from the order or plan, such as the commander's intent, CCIRs, task organization, situation, CONOPS, execution (including branches and sequels), and planning support tools (synchronization matrix, joint intelligence preparation of the operational environment products, and so on). It should describe the interaction of the elements of power to achieve the commander's visualization of the campaign. If the plan is approved as an operation plan, it must be reviewed periodically—every six months as required by Adaptive Planning and Execution, or when directed by the commander as his vision changes through his continual design approach to his theater.

Appendix A
Consolidated List of Unconventional Warfare Component Capabilities

USSOCOM component UW capabilities are specified and amplified in USSOCOM Directive 525-89. The directive conveys guidance for components to develop and maintain capabilities with specific application to UW that are aligned with the component's operational purpose, organizational design, competencies, and assigned responsibilities and functions (missions and SOF primary, secondary, and supporting core tasks) identified within USSOCOM Directive 10-1.

A component UW capability articulates the component's ability to execute a set of UW-specific tasks in denied, hostile, and politically sensitive environments, satisfying its core tasks. A component derivative UW capability articulates the component's potential ability to provide limited, derivative support through the application of an existing capability, function, expertise, or skill to the execution of a specific-UW task within specific operational environments (derived from the component's existing capabilities to execute its primary, secondary, or support core tasks).

U.S. ARMY SPECIAL OPERATIONS COMMAND UNCONVENTIONAL WARFARE-SPECIFIC ASSIGNMENTS

A-1. USASOC serves as the USSOCOM lead component in all matters pertaining to SOF conduct of UW. USASOC provides support of GCC/JFC/JTF/SOF HQ executing UW as directed by Commander, USSOCOM. It executes UW as a primary core task.

ARMY COMPONENT UNCONVENTIONAL WARFARE CAPABILITIES

A-2. USASOC has the UW capabilities described in Figure A-1, pages A-1 and A-2.

1.	Plan, coordinate, and exercise command and control of the military aspects of an unconventional warfare campaign across a distributed joint special operations area with varied operational environments and required modes of operation.
2.	Support/participate in compartmented unconventional warfare campaign planning in accordance with national- and theater-level requirements.
3.	Deploy personnel to target areas on a sustained basis, under any operational mode, and in any operational environment.
4.	Conduct time-sensitive area assessments within the joint special operations area.
5.	Provide forces capable of conducting unilateral, bilateral, or multilateral preparation of the environment operations.
6.	Plan and conduct intelligence operations in support of U.S. unconventional warfare and opposition elements.

Figure A-1. Army component unconventional warfare capabilities

Appendix A

7. Conduct strategic communications in support of unconventional warfare operations.
   a. Conduct information operations in support of unconventional warfare operations.
      (1) Conduct Military Information Support Operations in support of unconventional warfare operations.
      (2) Coordinate and conduct computer network operations in support of unconventional warfare operations.
         (a) Coordinate and conduct operations on the global information grid in support of unconventional warfare operations.
         (b) Leverage the cyberspace domain for the conduct of unconventional warfare operations.
      (3) Conduct electronic warfare in support of unconventional warfare operations.
      (4) Conduct operational security in support of unconventional warfare operations.
      (5) Conduct military deception in support of unconventional warfare operations.
   b. Conduct public affairs in support of unconventional warfare operations.
   c. Conduct defense support to public diplomacy in support of unconventional warfare operations.
   d. Conduct civil-military operations/Civil Affairs operations in support of unconventional warfare operations.
8. Provide and/or operate platforms to infiltrate/exfiltrate, resupply, and/or transport personnel and material by aircraft under any operational mode and in any operational environment.
9. Provide and/or operate platforms to infiltrate/exfiltrate, resupply, and/or transport personnel and material by ground, under any operational mode, and in any operational environment.
10. Locate and recover personnel and equipment within and from the joint special operations area.
11. Sustain unconventional warfare operations.
12. Communicate with the joint force and opposition to, from, and within the joint special operations area.
13. Provide or coordinate joint fires.
14. Develop, use, and coordinate cover support in compliance with the Defense Cover Program.
15. Coordinate Defense Sensitive Support.
16. Train, advise, and assist the opposition to develop an area command in a distributed joint special operations area, under any operational mode and in any operational environment.
17. Train, advise, and assist the opposition to recruit, develop, equip, employ, and transition through the development of an underground, auxiliary, and guerrilla force, in a distributed joint special operations area, under any operational mode and in any operational environment.

Figure A-1. Army component unconventional warfare capabilities (continued)

ARMY COMPONENT DERIVATIVE UNCONVENTIONAL WARFARE CAPABILITIES

A-3. USASOC has the derivative UW capability described in Figure A-2.

Provide and/or operate platforms to infiltrate/exfiltrate, resupply, and/or transport personnel and material by maritime vessel under any operational mode and in any operational environment.

Figure A-2. Army component derivative unconventional warfare capabilities

NAVAL SPECIAL WARFARE UNCONVENTIONAL WARAFRE-SPECIFIC ASSIGNMENTS

A-4. NAVSPECWARCOM serves as a USSOCOM supporting component for the maritime, riverine, and littoral aspects of UW. NAVSPECWARCOM provides support of GCC/JFC/JTF/SOF HQ executing UW as directed by Commander, USSOCOM. NAVSEPCWARCOM supports the maritime, riverine, and littoral aspects of UW as a secondary core task.

Consolidated List of Unconventional Warfare
Component Capabilities

NAVY COMPONENT UNCONVENTIONAL CAPABILITIES

A-5. NAVSPECWARCOM has the UW capabilities described in Figure A-3.

1. Provide and/or operate platforms to infiltrate/exfiltrate, resupply, and/or transport personnel and material by maritime vessel (surface/subsurface) under any operational mode and in any operational environment.
2. Train, advise, and assist the opposition to recruit, develop, equip, employ, and transition through the development of a maritime, riverine, and littoral-focused underground, auxiliary, and guerrilla force, in a distributed joint special operations area, under any operational mode and in any operational environment.

Figure A-3. Navy component unconventional warfare capabilities

NAVY COMPONENT DERIVATIVE UNCONVENTIONAL WARFARE CAPABILITIES

A-6. NAVSPECWARCOM has the derivative UW capabilities described in Figure A-4.

1. Coordinate, plan, and exercise command and control of special operations forces maritime, riverine, and littoral forces in support of an unconventional warfare campaign.
2. Support/participate in compartmented unconventional warfare campaign planning in accordance with national- and theater-level requirements.
3. Deploy personnel to target areas on a sustained basis, under any operational mode, and in any operational environment.
4. Conduct time-sensitive maritime, riverine, and littoral-related area assessments within the joint special operations area.
5. Provide forces capable of conducting unilateral, bilateral, or multilateral preparation of the environment operations.
6. Plan and conduct maritime, riverine, and littoral-related intelligence operations in support of U.S. unconventional warfare and opposition elements.
7. Infiltrate/exfiltrate, resupply, and/or transport personnel and material by ground, under any operational mode, and in any operational environment.
8. Locate and recover personnel and equipment within and from the joint special operations area.
9. Conduct strategic communications in support of unconventional warfare operations.
    a. Conduct information operations in support of unconventional warfare operations.
        (1) Coordinate and conduct computer network operations in support of unconventional warfare operations.
            (a) Coordinate and conduct operations on the global information grid in support of unconventional warfare operations.
            (b) Leverage the cyberspace domain in support of unconventional warfare operations.
        (2) Conduct electronic warfare in support of unconventional warfare operations.
        (3) Conduct operational security in support of unconventional warfare operations.
        (4) Conduct military deception in support of unconventional warfare operations.
    b. Conduct public affairs in support of unconventional warfare operations.
    c. Conduct defense support to public diplomacy in support of unconventional warfare operations.
10. Communicate with the joint force and opposition maritime, riverine, and littoral elements to, from, and within the joint special operations area.
11. Coordinate Defense Sensitive Support.
12. Develop, use, and coordinate cover support in compliance with the Defense Cover Program.

Figure A-4. Navy component derivative unconventional warfare capabilities

Appendix A

AIR FORCE SPECIAL OPERATIONS COMMAND UNCONVENTIONAL WARFARE-SPECIFIC ASSIGNMENTS

A-7. AFSOC serves as a USSOCOM supporting component and senior advisor for aviation aspects of UW. AFSCO provides support of GCC/JFC/JTF/SOF HQ executing UW as directed by Commander, USSOCOM. AFSOC supports the aviation aspects of UW as a secondary core task.

AIR COMPONENT UNCONVENTIONAL WARFARE CAPABILITIES

A-8. AFSOC has the UW capabilities described in Figure A-5.

1. Infiltrate/exfiltrate, resupply, and/or transport personnel and materiel by aircraft, and provide aviation command and control, under any operational mode, and in any operational environment. Currently, there is no ability to decontaminate aircraft exposed to a chemical, biological, radiological, or nuclear environment.
2. Assess, train, advise, and assist the opposition to recruit, develop, facilitate the transfer of equipment, employ, and transition through the development of an aviation-focused underground, auxiliary, and guerrilla force, in a distributed joint special operations area, under any operational mode, and in any operational environment. Currently, there is no ability to decontaminate aircraft exposed to a chemical, biological, radiological, or nuclear environment.

Figure A-5. Air component unconventional warfare capabilities

AIR COMPONENT DERIVATIVE UNCONVENTIONAL WARFARE CAPABILITIES

A-9. AFSOC has the derivative UW capabilities described in Figure A-6, pages A-4 and A-5.

1. Coordinate, plan, and exercise command and control of special operations forces aviation in support of an unconventional warfare campaign.
2. Support/participate in compartmented unconventional warfare campaign planning in accordance with national- and theater-level requirements.
3. Deploy personnel and equipment to target areas on a sustained basis, under any operational mode, and in any operational environment (including chemical, biological, and radiological).
4. Conduct time-sensitive aviation-related assessments within the joint special operations area.
5. Plan for and conduct airborne intelligence, surveillance, and reconnaissance operations in support of U.S. unconventional warfare and opposition elements, under any operational mode, and any operational environment.
6. Conduct strategic communications in support of unconventional warfare operations.
    a. Conduct information operations in support of unconventional warfare operations.
        (1) Coordinate computer network operations in support of unconventional warfare operations.
            (a) Coordinate operations on the global information grid in support of unconventional warfare operations.
            (b) Leverage the cyberspace domain in support of unconventional warfare operations.
        (2) Distribute Military Information Support Operations products via overt airborne platforms in a hostile environment.
        (3) Conduct electronic warfare in support of unconventional warfare operations.
        (4) Conduct operational security in support of unconventional warfare operations.
        (5) Conduct military deception in support of unconventional warfare operations.
    b. Conduct public affairs in support of unconventional warfare operations.
    c. Conduct defense support to public diplomacy in support of unconventional warfare operations.

Figure A-6. Air component derivative unconventional warfare capabilities

Consolidated List of Unconventional Warfare Component Capabilities

7.	Locate and recover personnel and equipment within and from the joint special operations area via overt airframes and ground recovery teams in any operational environment.
8.	Communicate with joint force and the opposition aviation elements, assets, and ground recovery teams to, from, and within the joint special operations area, using clandestine and traditional methods.
9.	Plan, coordinate, or conduct joint targeting and fires in any operational environment in support of clandestine and overt operations.
10.	Coordinate Defense Sensitive Support.
11.	Develop, use, and coordinate cover support in compliance with the Defense Cover Program.
12.	Plan, coordinate, and establish support for an unconventional assisted recovery coordination cell.

Figure A-6. Air component derivative unconventional warfare capabilities (continued)

MARINE CORPS FORCES SPECIAL OPERATIONS COMMAND UNCONVENTIONAL WARFARE-SPECIFIC ASSIGNMENTS

A-10. MARSOC serves as a USSOCOM supporting component for UW. MARSOC provides support of GCC/JFC/JTF/SOF HQ executing UW as directed by Commander, USSOCOM. MARSOC supports UW as a secondary core task.

MARINE COMPONENT UNCONVENTIONAL WARFARE CAPABILITIES

A-11. MARSOC has the derivative UW capabilities described in Figure A-7.

1.	Support/participate in compartmented unconventional warfare campaign planning in accordance with national- and theater-level requirements.
2.	Deploy personnel to target areas on a sustained basis, under any operational mode, and in any operational environment.
3.	Conduct time-sensitive area assessments within the joint special operations area.
4.	Provide forces capable of conducting unilateral, bilateral, or multilateral preparation of the environment operations.
5.	Plan and conduct intelligence operations in support of U.S. unconventional warfare and opposition elements.
6.	Conduct strategic communications in support of unconventional warfare operations.
	a. Conduct information operations in support of unconventional warfare operations.
	(1) Conduct electronic warfare in support of unconventional warfare operations.
	(2) Conduct operational security in support of unconventional warfare operations.
	(3) Conduct military deception in support of unconventional warfare operations.
	b. Conduct public affairs in support of unconventional warfare operations.
	c. Conduct defense support to public diplomacy in support of unconventional warfare operations.
7.	Infiltrate/exfiltrate, resupply, and/or transport personnel and material by ground, under any operational mode.
8.	Infiltrate/exfiltrate, resupply, and/or transport personnel and material by maritime vessel, under any operational mode.
9.	Communicate with the joint force and opposition guerrilla forces to, from, and within the joint special operations area.
10.	Train, organize, equip, and combat-advise opposition guerrilla forces in any operational environment.

Figure A-7. Marine component unconventional warfare capabilities

Appendix B

A 21st Century "Lawrence of Arabia": Attributes of the Special Forces Soldier

Do not try to do too much with your own hands. Better the Arabs do it tolerably than that you do it perfectly. It is their war, and you are to help them, not to win it for them. Actually, also, under the very odd conditions of Arabia, your practical work will not be as good as, perhaps, you think it is.

T.E. Lawrence, *27 Articles* (1917)

U.S. Army SF was designed to conduct UW, and the timeless attributes necessary to conduct UW are exactly the attributes most valued in the 21st century SF Soldier. The essence of Special Forces' UW excellence is comprised of those traits and skills which permit effective action through and with indigenous actors. Below is a list of the most common attributes required for any effective practitioner of UW.

REGIONAL ORIENTATION

B-1. Each SF group is oriented to a specific region of the world, which is within the AOR assigned by the Unified Command Plan to the geographically apportioned combatant commands. These commands include United States Central Command (USCENTCOM), United States European Command (USEUCOM), United States Pacific Command (USPACOM), United States Southern Command (USSOUTHCOM), and United States Africa Command (USAFRICOM). Each SF group orients toward specific areas of responsibility. Within each group, individual battalions and operational detachments have an even better defined focus within their areas of responsibility. The primary purpose for this regional orientation is to allow SF Soldiers to develop genuine expertise in the languages and cultures found in the assigned regional area. Such genuine expertise is the career goal toward which every "aspiring Lawrence" strives.

The beginning and ending of the secret of handling Arabs is unremitting study of them. Keep always on your guard; never say an unnecessary thing: watch yourself and your companions all the time; hear all that passes, search out what is going on beneath the surface, read their characters, discover their tastes and their weaknesses, and keep everything you find out to yourself. Bury yourself in Arab circles, have no interests and no ideas except the work in hand, so that your brain is saturated with one thing only, and you realize your part deeply enough to avoid the little slips that would counteract the painful work of weeks. Your success will be proportioned to the amount of mental effort you devote to it.

T.E Lawrence, *27 Articles* (1917)

B-2. Orientation helps commanders focus their personnel and training toward the peculiarities of the region. Examples are language training, cultural familiarization, cross-cultural communication, unique movement techniques, and military skills for a particular terrain and climate. Area orientation begins with formal qualification training in the different phases of the SF Qualification Course and is reinforced through continued training, repeated deployments to an assigned AOR, and a methodical career program of professional self-study and progressive language fluency. However, regional orientation does not preclude deployments to other areas of responsibility, which are usually based on mission requirements. SF's ability to adapt to and respect wide and varied cultures enables them to operate immediately where needed.

Appendix B

UNCONVENTIONAL WARFARE COMPETENCIES

B-3. SF Soldiers develop and strive to perfect the following core Soldier competencies. While these attributes are useful to any mission assigned, each attribute is important to effectively conducting UW. The quiet professional best able to develop competence in and integrate all of these attributes will be the best prepared to succeed in UW.

POLITICAL AWARENESS

B-4. SF Soldiers must understand U.S. policies, goals, and objectives and be able to articulate them in a manner that convinces foreign counterparts to support them. Similarly, they must understand the political context within which their counterparts operate. Political awareness is developed partially through staying current on U.S. policy documents and doctrine, partially through staying abreast of current affairs and maintaining a healthy interest in world and area affairs, and partially through intelligence preparation of the battlefield. Expert political awareness in UW must also include significant periods of time immersed in and studying the indigenous environment.

PROBLEM SOLVING

B-5. SF Soldiers analyze a situation, then adapt and apply U.S. doctrine; tactics, techniques, and procedures; equipment; and methods in a culturally sensitive and suitable manner to resolve difficult issues in nonstandard situations. Problem solving is not unique to UW. However, given that the typical SF element is often relatively isolated from most systems of outside support, the unconventional warrior must be self-reliant, conceptually agile, and capable of crafting innovative solutions to irregular threats and unusual problems. UW cannot be successfully waged by mindlessly following orders, checklists, or cookie-cutter solutions.

INTERCULTURAL COMMUNICATIONS

B-6. SF Soldiers influence foreign audiences by managing the content and flow of information through effective intercultural communications fundamental to mission success. Beginning with the SF Qualification Course, SF competency in intercultural communications rests on four pillars: interpersonal skills, nonverbal skills, language proficiency, and area and cultural orientation.

Interpersonal Skills

B-7. SF Soldiers listen with understanding, maintain an open mind, and observe and grasp the essential components of a given situation. They overcome ethnocentricity and treat foreigners as equals, communicating and teaching across intercultural barriers. They use interpersonal skills to get the desired action from a foreign counterpart.

Nonverbal Skills

B-8. Understanding intercultural dimensions of nonverbal gestures and behaviors is half of all human communication. Nonverbal communication requires an understanding of the gestures applicable to each culture. The SF Soldier's regional orientation permits him to focus on the nonverbal "vocabulary" of a specific region and to be sensitive to nonverbal communication.

Language Proficiency

B-9. SF Soldiers must achieve an established Defense Language Proficiency Test (DLPT) rating and demonstrate language proficiency to be qualified in their military occupational specialty (MOS). SF Soldiers continue to improve their language skills through unit-sponsored language training, repeated deployments into the region of orientation, and self-study. The motivation, self-discipline, and curiosity demonstrated by those who gain and maintain at least a modest working fluency with a target language is a key indicator of Soldier suitability and aptitude to conduct UW. Conversely, a demonstrated lack of such motivation, self-discipline, and curiosity is a key indicator of unsuitability and lack of aptitude for UW.

> **Foreign Language Proficiency**
>
> It is impossible to overstate the importance of foreign language proficiency as a key competency to conduct UW.
>
> UW is fundamentally concerned with persuading foreign partners to act in concert with U.S. objectives. The emphasis in UW is not to do oneself but to persuade the other to do on one's behalf. Sometimes, the only U.S. instrument available for gaining another's cooperation is the lone Soldier in a remote locale attempting to persuade the local people. Facility with foreign languages is a necessary skill to conduct such persuasion. Language ability is a tool to gain another's confidence. A demonstrated familiarity with their indigenous culture as it can sometimes only be conveyed through language, and the effort the Soldier has demonstrated to learn their language and customs, builds trust.
>
> A long-retired U.S. general once said, "We (Special Forces) need to speak 5.56." True. Unconventional warfare is "warfare." But if a Soldier does not also speak foreign languages, he is an inferior operative to conduct the quintessential skill of human persuasion upon which UW is based.
>
> Whoever does not understand this truth does not understand UW and is not competent to conduct it.

Area and Cultural Orientation

B-10. SF Soldiers develop a thorough understanding of the cultural and religious history, and the social, political, and economic dynamics of the population. Their understanding of the operational area extends to the physical factors of geography and climate within the specified region. Formal training and cultural immersion on deployments develops this orientation and understanding. SF Soldiers recognize the importance of, and cultivate, personal and professional contacts within their region.

PHYSICAL FITNESS

B-11. Physical fitness is not unique to UW or to SF Soldiers. However, SF Soldiers maintain the highest level of fitness because they understand that due to the often austere nature of special operations, their lives and mission success depend on the highest conditioning. In no other mission is this truer than in the conduct of UW where small teams can be isolated in harsh conditions beyond ready access to support networks for months or years.

B-12. Physical fitness for UW is measured by the ability to carry out mission-related physical events in extreme conditions while remaining mentally sharp. Physical confidence and stamina prevent timidity and indecisiveness when situations or conditions change and allow SF Soldiers to lead by example when working with indigenous partner forces. A high level of physical fitness also creates the opportunity for surprise by allowing guerrilla and other forces to attack in a manner or place the enemy did not think physically possible or feasible.

WARFIGHTING

B-13. SF Soldiers are tactically competent and have advanced training in operations, intelligence, medical skills, engineering, communications, and weapons. The SFODA is fully versed in light-infantry tactics, techniques, and procedures through battalion level. Their patrolling skills—carried out in unilateral, combined, or joint operations—include all aspects of combat patrolling. They also have a working knowledge of the employment of MISO and CA. SF practitioners of UW know how to integrate a wide range of state-of-the-art technology with a deep reservoir of ancient tactics and stratagems.

Appendix B

DEVELOPMENT OF HUMAN INFRASTRUCTURE

B-14. Human infrastructure is an invaluable contributor to a variety of SF missions, especially to the conduct of UW. More information relative to human infrastructure is in TC 31-16.

TRAINING

B-15. Long before most insurgencies are strong enough to wrest control of an opposing regime's power, they must organize, train, and expand. SF Soldiers provide a crucial cadre to assist and lead in this buildup. Each Soldier is individually proficient, both technically and tactically. Each is fully competent to assess unit and individual requirements, develop and implement programs to address identified needs, and evaluate the results of those programs. Each team member has specialized expertise and has cross-trained on other skills, giving each team a broad range of knowledge and capabilities to impart. The proficient unconventional warrior must be a competent and knowledgeable instructor.

JOINT, INTERAGENCY, INTERGOVERNMENTAL, AND MULTINATIONAL OPERATIONS

B-16. UW is inherently an operation involving multiple agencies, organizations, and stakeholders, and operatives must be able to facilitate JIIM operations. UW practitioners must understand the requirements of other agencies, Services, or nations to accomplish the mission. Army SF routinely operate as part of a joint, combined, and multinational special operations team under the guidance of civilian agencies, and are well prepared to support any combination of operations. Proactive liaison is essential to successful integration.

CLANDESTINE INFILTRATION AND EXFILTRATION

B-17. UW and U.S. support to resistance movements or insurgencies is typically very sensitive, requiring recurring yet discrete passage of borders and boundaries. SF Soldiers maintain proficiency in a wide variety of low-visibility and clandestine infiltration and exfiltration techniques. Although SF is a ground-oriented force, it is competent in air, sea, and land infiltration and exfiltration methods using any suitable and feasible modes of transportation. Unconventional warriors must also be able to adapt and exploit indigenous modes and opportunities to infiltrate and exfiltrate those indigenous personnel that they advise.

AUSTERE OR HOSTILE ENVIRONMENTS

B-18. Insurgencies and resistance movements usually have to withdraw into remote, inaccessible, inhospitable, and undesirable areas to survive—particularly in the early phases of resistance. UW operatives must maintain the ability to operate for extended periods in such hostile, remote, and austere environments with little or no external support. SF Soldiers know that the ability to operate in these challenging environments requires mental as well as physical preparedness. Mental toughness, physical endurance, tenacity, and perseverance are necessary qualities of the unconventional warrior.

ADVANCED TECHNOLOGY

B-19. It is an individual responsibility to identify areas where technologies, civilian or military, can enhance mission accomplishment. SF Soldiers stay abreast of all commercial off-the-shelf advances and innovative applications in their assigned areas. They apply creative solutions to mission requirements through the innovative use of available technology coupled with nonstandard procurement procedures. SF Soldiers understand that UW is always fundamentally about human interaction, and applying technology to such interaction is an aid to UW; it does not make it obsolete. Given the current advanced state and widespread availability of personal communication devices, UW practitioners must anticipate and leverage this technology to conduct successful IO.

> *All men dream; but not equally. Those who dream by night in the dusty recesses of their minds wake in the day to find that it was vanity: but the dreamers of the day are dangerous men, for they may act their dreams with open eyes, to make it possible. This I did.*
>
> T.E. Lawrence, *Seven Pillars of Wisdom* (1922)

Appendix C
Selected Examples of U.S.-Sponsored Unconventional Warfare Activities

"U.S.-sponsored" UW is the current, modern understanding of what UW means. This understanding is based on the idea that the USG uses UW as a strategic policy option by supporting foreign actors. The United States has conducted such activities since WWII. Although one cannot talk of "U.S.-sponsored" UW in the modern sense when such activities are conducted during the American Revolution or American Civil War, antecedent examples of UW activities contributing to today's concept are noted.

AMERICAN REVOLUTIONARY WAR

C-1. Colonial patriots engaged in underground activities when they established the "Committees of Correspondence" in the prelude to overt battle in the American Revolution. The committees functioned as organizational and propaganda cells which coordinated anti-British activities and fomented popular resentment against British authority. Prior to Lexington and Concord, the committees had become the nucleus for a shadow government around which revolutionary activities coalesced. The committees would also function as the first organized bodies that would eventually make up the Continental Congress.

C-2. UW activities the committees engaged in included—
- Proving organizational leadership.
- Slanting news to favor their cause and disparage the regime.
- Identifying supporters and preparing them for leadership positions.
- Identifying opponents and collaborators and arranging punishments and humiliations for them.
- Arranging boycotts and other forms of civil disobedience.
- Popularizing and proselytizing American national identity as distinct from the British government.
- Supervising elections for patriot government bodies and so on.

C-3. From the military perspective, General Washington also exploited underground and auxiliary-type methods by establishing a widespread intelligence network to spy on British movements and intentions. Of special note in the Carolinas was Lieutenant Colonel Francis Marion, "The Swamp Fox" (later promoted to Brigadier General). He commanded a disparate, rugged band of Continentals and used guerrilla tactics honed while fighting the Cherokee in the French and Indian Wars. Seeking cover in swamps and taking advantage of low-country riverine systems, he and his troops harassed the British with a success out of all proportion to their small numbers.

AMERICAN CIVIL WAR

C-4. In the decades leading up to overt major operations in the American Civil War, mostly northern abolitionist and church groups ran a kind of "evasion and escape" network called the "Underground Railroad." The so-called "railroad" was a network of escape routes and safe houses run by an "underground" collection of guides, handlers, front men, and financiers that helped runaway slaves escape from the American south and find sanctuary in the northern Free states, Canada, or Mexico. Although exact figures are in dispute, tens of thousands may have evaded using this UW activity.

C-5. Guerrilla fighters—particularly cavalry—operated behind the lines on both sides of the war. The most famous guerrilla fighter was Confederate Colonel John Singleton Mosby who ranged the northern

Appendix C

and western counties of Virginia. Following tactics and strategy perpetuated in the Carolinas during the Revolutionary War, he and his command were successful in spreading disinformation to Federal forces about troop strength and positions, disrupting Federal supply and communication lines, and distracting large contingents of troops to protect their rear areas. He and his band were the scourge of Union generals. Other behind-the-lines partisan raiders such as the pro-Confederate William Quantrill or the pro-Union Redlegs and Jayhawkers were sometimes effective against enemy troops but were often regarded as little more than brigands and "bushwhackers" who terrorized the population of Missouri and Kansas.

WORLD WAR II OFFICE OF STRATEGIC SERVICES

C-6. The Office of Strategic Services was the product of Colonel William J. Donovan (later promoted to Major General). A veteran of World War I, he had received the Medal of Honor for his actions as an infantry commander on the Western Front in October1918. When WWII finally erupted in Europe and threatened to engulf the United States, Donovan became convinced that a new type of organization would have to be formed to collect intelligence and wage secret operations behind lines. In 1941, President Roosevelt directed Colonel Donovan to form a quasi-civilian agency named the Office of the Coordinator of Information. In 1942, the agency was activated as a military unit renamed the Office of Strategic Services.

C-7. The sudden U.S. involvement in WWII warranted a still more unconventional kind of warfare which focused on the development of indigenous fighting units to combat enemy forces. This UW became a primary concentration of the agency. Special teams were formed to provide liaison and logistics support, and to advise and assist resistance groups in the conduct of small-unit guerrilla warfare. Recruited heavily from ethnic communities in the United States, operatives were specially chosen and trained to perform defined missions and work hand-in-hand with specific partisan groups in Europe and East Asia. To ensure volunteers were up to the task, the Office of Strategic Services employed a detailed and comprehensive selection and assessment program as part of their recruitment process.

C-8. These personnel with extensive cultural and language expertise were trained in both conventional and guerrilla tactics and a dominant feature of training was their detailed planning and preparation for missions. All the features of the Office of Strategic Services' doctrine, training, organization, and operations comprised the original model and template for the later development of U.S. Army SF. More than any one entity or antecedent, the Office of Strategic Services special force units formed the later U.S. Army SF and their focus on UW. The two distinct types of units formed for operations in Europe were the Operational Groups and the Jedburgh teams. The Operational Groups would also conduct operations in the China-Burma-India Theater.

OFFICE OF STRATEGIC SERVICES OPERATIONAL GROUPS

C-9. The actual size of Operational Group teams in the field would vary greatly, but they were essentially 30- to 34-man units (26 to 30 enlisted men and 4 officers) who generally fought as split teams or cells of 15 to 17 men. Primarily infantrymen and demolition experts, these teams also contained medical technicians and radio operators. The teams inserted behind enemy lines by parachute or by boat to conduct guerrilla operations, conduct raids, lead and train the local resistance movements, and generally create the maximum amount of confusion for the enemy. By mission and design, the Operational Groups were the most direct pattern for the SF "A" team, and the 12-man operational detachment is a faithful modification of the Operational Groups concept.

JEDBURGH TEAMS

C-10. In 1944, the Jedburgh Mission was a major operation of the Office of Strategic Services in Europe. The 93 three-man Jedburgh teams typically consisted of a French officer or an individual from the country in which the team was to operate, an American or British officer, and an enlisted radio operator. They received training in close combat, infiltration and exfiltration techniques, small-unit tactics, light weapons, demolitions, and a host of skills necessary to survive behind German lines. Formed to deploy in support of the D-day Normandy invasion and subsequent push toward Berlin, they parachuted into France, Belgium, and the Netherlands and performed as liaison for resistance groups. They supplied, trained, and guided partisans in attacks on German units and communications; ambushed retreating columns; kept major installations from demolition; rescued downed Allied pilots; screened the flanks of conventional forces; and gathered vital intelligence.

DETACHMENT 101

C-11. In 1943, Office of Strategic Services Detachment 101 was developed under Colonel Carl Eiffler to conduct operations in the China-Burma-India Theater. Responsible to not only recruit and train, but also to command indigenous members against a much greater Japanese force, Office of Strategic Services Detachment 101 was the first unit ever to be formally built for this purpose. They ran disinformation campaigns, gathered intelligence, performed sabotage, scouted for conventional units, and rescued downed Allied flyers. They had their own aviation assets, a signal company which included a company of carrier pigeons to relay messages, and a medical detachment. They later spawned other organizations, including Detachment 202 in China, Detachment 303 in Thailand, and Detachment 404 in French Indo-China.

GUERRILLAS IN THE PHILIPPINES

C-12. General Douglas MacArthur would not allow the Office of Strategic Services entry into his Southwest Pacific Command until very late in WWII. But he is acknowledged to have been generally receptive to special operations. Among his special operations forces were the Alamo Scouts and extensive networks of guerrilla forces throughout the Philippine Islands. The two best known and most effective guerrilla organizations were Wendell Fertig's on Mindanao and Russell Volkmann's on Luzon.

C-13. After the U.S. forces in and around the island of Luzon surrendered in May 1942, Fertig decided not to give himself up to the Japanese. Colonel Fertig slowly assumed control over the existing rival guerrilla units on Mindanao and turned them to harassing the Japanese. Over the next two and a half years, Fertig created and commanded the Mindanao segment of the United States Forces in the Philippines (USFIP), recruiting escaped prisoners-of-war, and soldiers and American civilians who had refused to surrender. His best resource for his leadership cadre was American servicemen who had either not surrendered or had escaped from prisoner-of-war camps. Many were already fighting as guerrilla leaders. However, his real strength came from the thousands of patriotic Filipinos, many of whom were enraged by Japanese atrocities, and who now joined existing or formed new organizations under his command.

C-14. In addition to ambushing, raiding, and generally harassing occupying Japanese forces, Filipino guerrillas provided valuable intelligence reports to Allied strategists. Coast-watching networks played a significant role in the American naval victories of the Philippine Sea and Leyte Gulf. Downed aviators were rescued and helped to evade through an evasion network that included exfiltration by submarine. Fertig even created a "navy" by arming several small merchant vessels which he used to protect convoys of small vessels that helped distribute supplies and which occasionally attacked inland Japanese patrol barges. In preparation for the Allied invasion, Fertig's guerrillas constructed clandestine airstrips in the jungle. After invasion began, the Mindanao guerrillas acted as guides and seized key terrain to aid the advance.

C-15. By late 1944, Fertig commanded a resistance organization that numbered tens of thousands, and through providing hope, channeling resistance fervor into organized effort, and distributing supplies and medicines, held the allegiance of hundreds of thousands of civilians. Over three years of resistance, the Japanese had been forced to relinquish most of the island, garrisoning only the coastal cities with troops badly needed elsewhere.

C-16. On the northern main island of Luzon, Colonel Volkmann led a similar effort. At the outset of occupation, the resistance was badly fractured and most American guerrilla leaders were ineffective. However, as Japanese brutality inspired the locals to resist, and as a series of less-competent American leaders were captured and executed, Volkmann eventually built an effective resistance organization. Initially, Volkmann concentrated on the development and buildup of the organization and establishment of an intelligence net, avoiding major clashes with the Japanese.

C-17. Volkmann used personal charisma and leadership to good effect. To gain public confidence and support, he brought rival tribes and factions together through personal diplomacy and instituted a crackdown against bandits who were looting and plundering the natives. Faced with an extensive Japanese network of spies and informers, he and his subordinates also launched a ruthless counterespionage campaign to eliminate the collaborators. Resistance operatives infiltrated the neighborhood associations and the constabulary to identify the informers. Within six months, those not executed had fled to the protection of Japanese garrisons. With the occupying power's network of informers eliminated, the resistance was able to organize and mobilize the villages and towns into an expanded anti-Japanese intelligence and sabotage effort.

C-18. As the overall resistance organization expanded, so too did the size and effectiveness of the guerrilla force. Like Fertig, Volkmann made good use of escaped and evading American officers, former Philippine Constabulary officers, and even local business officials for his guerrilla cadres. This cadre trained local guerrillas in guerrilla tactics, demolitions, and scouting. In addition to years of frustrating and tying down Japanese occupation forces, they eventually also played an important role in supporting the American invasion. Volkmann's guerrilla-guided U.S. forces harassed Japanese counter movements, recovered and assisted evading downed pilots, guarded captured areas, and eliminated bypassed enemy detachments and outposts that released badly needed U.S. troops for other duties.

POST-WORLD WAR II

C-19. If WWII provided the classic U.S. template of UW as support to major operations, the concept of UW as a stand-alone national strategic option was fully developed during the subsequent Cold War decades. In fact, the political warfare so central to the Cold War era was so influential to the concept of UW that some doubted its continued value after the fall of the Soviet Union. However, the continued utility of UW was demonstrated anew as various 21st century threats emerged.

CENTRAL INTELLIGENCE AGENCY AND U.S. ARMY SPECIAL FORCES

C-20. After WWII, effective 1 October 1945, President Truman disbanded the Office of Strategic Services. Its functions and assets were moved to other agencies in the State Department or were discarded. UW capabilities in the military rapidly degenerated and senior military leaders focused on the disposition of conventional forces to oppose the Soviet threat. In response to the growing danger posed by Communist aggression and subversion in Europe and Asia, the National Security Act of 1947 created the Central Intelligence Agency. The Central Intelligence Agency took the lead in UW and began coordinating intelligence overseas and conducting strategic covert activities.

C-21. Several of the Office of Strategic Services operatives were prominent and formative in the early days of the U.S. Army SF, and are recognized as among the Founding Fathers of the SF regiment. Colonels Aaron Bank and Carl Eiffler and Lieutenant Colonel Hubert Brucker were among former Office of Strategic Services members who worked to establish the U.S. Army SF in the late 1940s and early1950s. Lieutenant Colonel Russell Volckmann, noted for the conduct of guerrilla warfare in the Philippines, also used his experience and lessons learned to help formulate the doctrine of UW that became the cornerstone for U.S. Army SF. As the Cold War threat became more apparent and more serious, SF trained to execute both stay-behind and classic infiltration UW operational plans in the event of a communist overrun of Western Europe.

COLD WAR AND IMMEDIATE POST-COLD WAR

C-22. Compared to the successes during the "classic era" of UW—especially in the Pacific and China-Burma-Indian Theaters—the UW conducted during and immediately following the long Cold War provides a record of uneven results. The ability of U.S. Army SF to conduct widespread UW throughout Europe (and elsewhere) in the advent of a Soviet invasion must be counted as having been a strategically useful policy option. UW (or its component activities) was successful, if limited, in places like Tibet or Korea. However, other efforts at UW in the Baltic States, the Balkans, or Cuba were failures.

Korean Conflict

C-23. When the Korean Conflict began in 1950, many anticommunist displaced persons from North Korea traveled to nearby islands and began a covert war of their own against the North. The partisan force was organized in 1951 as the 8240th Army Unit. This United Nations Partisan Force, Korea operated in much the same manner as the guerrillas had done in the Philippines or as the European partisans had done under the Operational Groups and Jedburgh teams. Eventually attaining the rank of Brigadier General, Russell Volckmann had also become involved with the partisans until his evacuation in 1951. By 1952, members of the new 10th SF Group began to augment the cadre. They added evasion-and-escape operations to unit tasks, and dispatched several operatives to various known prisoner-of-war camps in North Korea. In March 1953, the 10th SF Group sent more Soldiers to the 8240th Army Unit, who eventually established the 8086th Army Unit and worked with partisan forces on their own. They were credited with effectively tying up more than 75,000 enemy troops by the end of the war.

C-24. Nevertheless, the resistance efforts were unable to sustain any pockets of resistance on the mainland for prolonged periods of time. The decision to disband the Office of Strategic Services had led to an atrophying of UW skills. The general lack of familiarity with these types of operations and the failure to exploit indigenous population groups contributed to the military's decision to reestablish a UW capability by creating U.S. Army SF.

Central Intelligence Agency in Albania and Latvia (1951–1955)

C-25. The Central Intelligence Agency used these opportunities to test theories for rolling back communist domination of Eastern Europe. These locations were chosen due to their relatively small size. Both efforts were failures. The occupying power and repressive local regimes had had time since WWII to solidify their repression making it difficult to begin a resistance movement. The essential underground and auxiliary never had realistic opportunities to develop into a serious threat.

Central Intelligence Agency in Tibet (1955–1970)

C-26. This effort was well-executed tactically and did delay Chinese victory. However, no amount of small-scale tactical success was going to prevent such victory forever. This effort is notable for two reasons. It demonstrates the ability to keep a cross-border shaping operation covert over a sustained period, and is an example of successfully exfiltrating an indigenous operative for training and then reinserting him back into the AO. This technique avoids the insertion of U.S. forces with the attendant increased risk of compromise.

Indonesia (1957–1958)

C-27. Fearing that the Indonesian Government was leaning toward communism, the Central Intelligence Agency supported several Indonesian military officers who claimed to control a rebel army made up of former military units. The agency supported this effort with foreign mercenary air forces, some of whom were captured and exposed by the Indonesian government. The resulting scandal forced the Central Intelligence Agency to abandon the effort. The rebel army had little popular support among the population while the Indonesian government still held control of the majority of its military. When the Central Intelligence Agency withdrew support, the Indonesian Government crushed the rebellion with conventional military forces. This case demonstrates the political risk involved with using foreign mercenaries. More importantly, it shows that externally supported rebellions without significant popular involvement and support are likely to fail.

Cuba and the Bay of Pigs (April 1961)

C-28. The Central Intelligence Agency made similar fatal errors in prosecuting UW activities in Cuba. Although there were some reasonably successful efforts at developing an underground, they were not sufficiently widespread or coordinated to organize a persistent popular resistance campaign. Instead, the "resistance movement" was generally "manufactured" from exiled paramilitary and mercenary forces, rather than from one developed organically in the AO.

C-29. The Central Intelligence Agency made other operational mistakes as well. They chose a quasi-conventional amphibious assault on Cuba—with assumed U.S. naval support—rather than on a popular action-oriented effort. USG decisionmaking and interagency coordination were poor. Intelligence on current conditions in Cuba was inadequate and misleading. The invasion site was not operationally or tactically sound. The U.S. naval support was eventually withheld. Much of the supporting equipment was lost or misdirected. Operational security was lost and the Cubans were alerted to the invasion force. The Bay of Pigs was complete military failure and political debacle.

C-30. The lessons demonstrated by the Cuban effort are numerous and significant. UW requires covert infrastructure and support mechanisms. UW campaign planners must understand the dynamics of insurgency and resistance operations; this must include the proper importance of the underground and popular political organization and preparation versus the relatively subordinate function of guerrilla warfare. There is no substitute for adequate accurate intelligence on popular resistance feasibility. Similarly, insurgent psychological operations, propaganda, subversion, and so on must be adequate before a decision by force-of-arms is attempted. Long-exiled participants—even if natives of the target country—are not a substitute for active insurgent operatives embedded within the population. Operational security is fundamental to successful unconventional campaign success. UW requires USG interagency coordination and cooperation and national strategic coherence and oversight.

The Central Intelligence Agency and Special Forces in Laos (1959–1962)

C-31. In 1960 and 1961, Army SF from 1st and 7th Groups collectively had 21 operational detachments serving as advisors to Laotian Army Battalions. This counterinsurgency effort was secret and predominantly in the southern Laos. Tribal elements were also trained to provide civil defense in some areas to complement the regular military. Concurrently, SF conducted other training in support of two UW initiatives to train Meo and Kha tribesmen to conduct guerrilla warfare in remote, rural tribal areas occupied by the North Vietnamese Army and forces of the Pathet Lao. The campaign envisioned the tribal guerrillas raiding staging areas and safe houses behind the lines to interdict the Ho Chi Minh trail running through Laos and Cambodia. Ultimately, full execution of these activities was never authorized.

Central Intelligence Agency and Special Forces in North Vietnam (1961–1964)

C-32. In 1960, the Central Intelligence Agency began attempts to establish agent networks within North Vietnam. The program was highly penetrated by double agents and never produced any viable results beyond a handful of questionable agents. The hopes of conducting UW against North Vietnam had diminished by 1964 when the military assumed control of the operation. The Military Assistance Command, Vietnam—Studies and Observation Group (MACV-SOG) actual name had originally been "Unconventional Warfare Task Force."

C-33. The SF continued the program by parachuting the questionable trainees over North Vietnam and terminating further support. In the teeth of the communist security posture, results from this program were predictable; none of the agents were heard from again. Despite widespread popular misunderstanding, virtually all SF activity in Vietnam through 1975 was counterinsurgency—not UW.

The Central Intelligence Agency and Special Forces in Nicaragua and Honduras (1980–1988)

C-34. The United States had success—if limited and temporary—in coercing and disrupting the Sandinista government's activities to foment and support revolutionary activities throughout Central America by "UW-like guerrilla activity." However, the actions of the so-called "contras" cannot be considered a good example of the UW concept. The Contras did conduct guerrilla warfare out of border safe havens in both Honduras and Costa Rica. However, they were essentially mercenaries with no popular base of support, underground or auxiliary structure, nor any realistic plans to develop any.

Central Intelligence Agency and Special Forces in Pakistan and Afghanistan (1980–1991)

C-35. Following the Soviet invasion of Afghanistan in 1979, the United States provided training and material support to mujahedeen fighters through allied intermediaries in Pakistan. The provision of lethal aid—particularly Stinger antiaircraft missiles—was useful to the cumulative attrition, and eventual withdrawal, of Soviet forces. These activities, which enabled a resistance with guerrilla, auxiliary, and underground components in denied territory to defeat and expel occupying forces, must be considered a successful instance of UW.

C-36. Nevertheless, this early Afghanistan example also provides a useful object lesson in second- and third-order effects and unintended consequences. The initial and immediate results of expelling and humiliating the Soviet Army were achieved. The moral imperative that international armed aggression should be resisted and overturned was upheld. Moreover, the defeat in Afghanistan was a contributing factor to the eventual demise of the Soviet Union.

C-37. However, the victory left a power vacuum that was ultimately filled by mortal foes of the United States. The mechanisms for mujahedeen support in Pakistan were now organized, empowered, and experienced. The mujahedeen and an entire generation of foreign observers were further radicalized by the conclusion that it was the irresistibility of their religious authority and radical worldview—not Soviet fatigue in the face of ancient guerrilla tactics combined with space-age technology inputs—which was responsible for their victory. Having defeated one pagan superpower, the mujahedeen now turned to attacking the remaining superpower in New York on 11 September 2001 and elsewhere.

C-38. This case provides a good example of the role and importance of neighboring allied countries' support bases to enable persistent covert operations. Most importantly, however, it highlights the importance of UW feasibility studies and the campaign planners' and senior decisionmakers' best judgment in considering the longer-term possible ramifications of conducting UW.

Kuwait (1990–1991)

C-39. Spontaneous resistance elements did arise among the Kuwaiti population after the invasion by Iraq. However, the overall campaign waged to decimate Iraqi forces was focused on decisive engagement by major operations, and no significant UW planning occurred. This "non-example example" illustrates that UW is often unnecessary, often overcome by the speed of events, and is usually unsuited as an off-the-shelf military "contingency response." Kuwait also provides an example of terrain largely unsuitable to support resistance activities.

Cold War Contingency Plans for Scandinavia and Europe (1952–1989)

C-40. Although not executed, it is worth noting that SF would have implemented UW contingency plans in the event of communist bloc invasion in Europe and elsewhere during the Cold War. The specifics of most of these contingency plans remain classified but the expected activities are consistent with the information in this publication.

POST-11 SEPTEMBER 2001 TO THE PRESENT

C-41. Events at the dawn of the new millennium demonstrate the continued utility of UW. Not only were UW campaigns conducted against post-9/11 terrorist-sponsoring nation-state threats, but UW continues to be a valuable national strategic policy option to coerce, disrupt, and—if necessary—overthrow nation-state opponents.

AFGHANISTAN—OPERATION ENDURING FREEDOM (2001–2002)

C-42. Between 19 October and 20 November 2001, Task Force Dagger (5th SF Group) inserted 11 operational detachments into Afghanistan. The detachments linked up with and coordinated the actions of loosely combined Tajik, Uzbek, Hazar, and Pashtun warrior bands referred to as the Northern Alliance. The Northern Alliance had been attempting to seize power from the Taliban government, which was

Appendix C

providing safe haven to Osama bin Laden. Prior to October, the Alliance had had limited success. U.S. SF brought space-age technology in support of tribesmen on horseback. These combined efforts were then overwhelmingly successful.

C-43. It is worth noting that the general weakness of the Taliban as a military force and the dramatic increase in lethal capability by the integration of U.S. air power created a state of parity between the Northern Alliance and Taliban not typical of UW efforts. Success was largely due to the operational capabilities of the inserted personnel—their access to overwhelming air superiority and ability to coordinate tactical actions through state-of-the-art telecommunications. Moreover, although U.S. forces did conduct "activities to enable a resistance to overthrow a government by operating with and through a guerrilla force in denied territory," the evidence of organized underground and auxiliary efforts is weak. Considering the power mismatch between a military superpower and a ragtag, incompetent Taliban army, and considering that (except for the very first pilot teams) once airstrikes began the Northern Alliance was operating along identifiable lines—not in "denied" territory—this case of U.S. Army SF conducting UW could be considered more correctly as the vanguard of a conventional invasion.

C-44. So, the initial stage of Operation ENDURING FREEDOM was "UW." However, in some ways it just barely satisfied the conditions of the definition. It was a case very far from the classic UW model, which typically assumes a relatively protracted period of development and very much assumes involvement of the populace far beyond the super empowerment of mere guerrilla bands. To identify that this case of UW is atypical—even an aberration—is not to deny the participants the honor they have earned. But UW campaign planners should not regard Operation ENDURING FREEDOM as a standard UW template.

IRAQ—OPERATION IRAQI FREEDOM (2002–2003)

C-45. As part of the invasion of Iraq in 2003, two UW campaigns were considered. A Northern effort with the Kurdish Resistance—predominantly made up from the Patriotic Union of Kurdistan (PUK) and the Kurdistan Democratic Party (KDP)—and a southern effort comprised of various Shia resistance groups. The Northern effort was intended to keep the Iraqi forces focused on the northern part of the country, making them unavailable to counterattack the coalition southern invasion. The Southern effort would have been executed to facilitate the introduction of invasion forces. Conventional commanders saw little or no need for working with the Shia resistance and, for a variety of reasons, the southern effort was never executed. The example of Iraq is significant because it demonstrates a scenario where two UW efforts could have been conducted in support of the conventional campaign, but for entirely different purposes.

C-46. The northern effort was considered the third priority for special operations efforts during the invasion of Iraq, until Turkey refused U.S. forces access to Turkish bases. This meant that the United States would not have a second conventional maneuver force entering Iraq from the north. Without additional U.S. conventional forces invading from Turkey, only six U.S. divisions would participate in the initial invasion from the south. The importance of sending U.S. SF into Northern Iraq to help mobilize Kurdish forces' participation then dramatically increased. Ultimately, approximately 30,000 Kurdish forces, supported by two battalions of Army SF, successfully tied down 12 of Iraq's 20 divisions during the invasion and liberated the cities of Kirkuk and Mosul.

C-47. Were the actions taken by SF in the earliest stages of Operation IRAQI FREEDOM UW? Like in Afghanistan, U.S. forces did "conduct activities that enabled an insurgent to disrupt and contribute to the overthrow of a government. " However, also like Afghanistan, the claim that U.S. forces were operating in "denied territory" is only valid in the narrowest interpretation of "within the boundaries of the sovereign state of Iraq." Practically speaking, the SF Soldiers air landed on secure airfields in a de facto independent Kurdistan which had been evacuated by Iraqi troops since 1991 when UN Security Council Resolution 688 had declared a "safe haven" for Kurds and when the United States and Britain had established a "No-Fly Zone" preventing Iraqi domination. Underground and auxiliary activities did occur, notably in the seizure of Kirkuk, in dissemination of PSYOP efforts and in intelligence collection. However, and again like Afghanistan, there was little demonstrated need for classic underground and auxiliary activities within unoccupied Kurdistan. Finally, the Kurds represented a very well-established and developed resistance force compared to the Northern Alliance. However, also like Afghanistan, the insertion of SF basically provided space-age technological super empowerment to standing indigenous coalition light infantry battalions fighting along traceable battle lines with distinct front and rear areas—hardly the classic concept of guerrilla forces.

C-48. Every instance of UW will be different and once again, to identify that this case of UW is atypical—even an aberration—is not to deny the participants the honor they have earned. But UW campaign planners should not regard Operation IRAQI FREEDOM as a standard UW template.

BEYOND TERRORISM AND DEEP INTO THE 21ST CENTURY

C-49. UW is not synonymous with just "overthrowing" governments. In fact, UW will progress to the overthrow of a government or expulsion of an occupying power only in rare cases. However, UW can also be used to "coerce or disrupt" government and occupying powers—a broad and flexible interpretation of employing U.S. power. UW remains a valuable strategic policy option for national and theater-level decisionmakers, and will continue to be so for the conceivable future. If any UW activities are currently being conducted, they would be classified and therefore inappropriate for discussion in this publication.

This page intentionally left blank.

Appendix D
Notable International Practitioners of Insurgent, Resistance, or Revolutionary Warfare Throughout History

UW as a modern U.S. DOD doctrinal concept of U.S.-supported insurgency traces its lineage, and draws its enduring conceptual inspiration, from a "classic period of UW" during WWII. This classic period—forged in the global struggle against Nazis, fascists, and Japanese Imperialists—was refined into its purest concept during the Cold War decades against global communism. Some believe that insurgency is only a modern phenomenon only formed in the middle 20th century, while others believe that the declared end of the Cold War signifies the modern irrelevance of insurgency or U.S.-sponsored support to insurgency. Both views are too historically shallow, and too U.S. and military-centric. Both views are mistaken.

The international examples surveyed below provide short historical descriptions of insurgency, resistance, or revolutionary warfare activities in action by other countries or actors. Others have used the techniques and activities common to UW with success for millennia. The student of UW—and planner of UW campaigns—can find historical insurgency, resistance, or revolutionary warfare activity precedents useful to understanding and developing future UW campaigns.

Note: Listing of techniques and activities used by other nations to illustrate insurgency, resistance, or revolutionary warfare does not denote nor imply that the USG uses or supports all of these techniques or activities.

SELECT HISTORICAL ANTECEDENTS OF TODAY'S UNCONVENTIONAL WARFARE ACTIVITIES

D-1. Throughout history, undergrounds have been the foundation and core of resistance and revolutionary movements, frequently directing political activities, performing many different kinds of organizational and operational functions, and, in many cases, supporting overt guerrilla warfare.

D-2. During the Peloponnesian War in the fifth century B.C., underground agents in Athens and Sparta tried to subvert the governments of each other's city states. Spartan agents working within democratic institutions of Athens were particularly effective in their propaganda and agitational activities which were designed to create internal dissension and develop popular distrust of the government.

D-3. In China, it early became common strategy for underground movements to form guerrilla bands and organize support groups among citizens in order to overthrow a local warlord or mandarin. During the Tang dynasty, in the ninth century A.D., a popular leader named Huang Ch'ao set himself up as a sort of Robin Hood, taking goods from the wealthy and from government officials, distributing them to starving peasants, and organizing the peasants behind him. The underground and guerrilla tactics of this era provided lessons which were absorbed by subsequent revolutionary groups and were ultimately reflected in the actions and writings of Mao Tse-tung.

D-4. In Europe during the Middle Ages, there were underground movements among underprivileged or persecuted groups. Among these were the Jacquerie, a loosely knit organization of beggars and thieves in 14th century France; the peasant groups in southern Germany; and the clandestine heretical religious sects such as the Albigensians in southern France.

D-5. Underground movements also played a role in numerous dynastic conflicts, such as the "War of the Sicilian Vespers." Around 1280, Charles of Anjou, the French King of the Two Sicilies, was preparing to invade the Byzantine Empire. The elderly Byzantine Emperor Michael VIII anxiously sought by all possible means to prevent Charles from carrying out his plans. Michael was aware of considerable unrest in Sicily which had resulted from Charles' harsh rule and was also aware of the frustrations of Spain's King of Aragon, Pedro III, who asserted claims to Sicily. After a promise of aid from Pedro, Michael's agents aggravated the unrest among the Sicilian people, and as a result of their underground activities a popular uprising in Sicily resulted in the massacre of 8,000 French troops in 1282. Soon after this, Pedro, pretending to lead a crusade to the Holy Land, changed course to land on the shores of Sicily and help the insurgents against Charles. The War of the Sicilian Vespers lasted for 20 years and eventually culminated in the Aragonese conquest of Sicily from the Angevins. The Sicilian spirit of independence, which first manifested itself during the War of the Sicilian Vespers, was to lead to further revolts against a variety of rulers in subsequent years. Though not a political independence movement, the Mafia developed from a group of outlaws who, for pay, protected landed estates, to an extralegal organization which strongly influenced the political activity of the island through economic means. To the present day, the Mafia claims to be the spearhead for the true interests of the Sicilian people.

D-6. The 14th century also provides some interesting cases of people resisting an outside invader and regaining national independence via movements involving undergrounds. For instance, after the second conquest of Scotland by Edward I of England, a Scottish resistance movement was begun in 1306 by Robert Bruce, a former vassal and ally of Edward. Bruce combined ambush and surprise attacks against English garrisons in Scotland with raids extending into the border regions of northern England. Using these tactics, Bruce had driven the English from most of Scotland by 1314 and was crowned Robert I, King of Scotland. In the same year, he defeated the invading army of Edward II, son of Edward I, at the Battle of Bannockburn and thus restored the independence of Scotland.

D-7. The Renaissance and Reformation were marked by the development of many groups with the primary goal of plotting political change. Perhaps the most notable of these were the numerous small clandestine groups planning coups throughout the major cities of central and northern Italy during the 15th and 16th centuries. Intrigue and assassination were common weapons in these struggles for wealth and power, aptly described by Machiavelli. The Counter Reformation saw the growth of international movements organized in the name of the Catholic and Protestant faiths, but dedicated to achieving political as well as religious goals. Elaborate intelligence and counterintelligence systems, propaganda campaigns, and other forms of political warfare aimed at influencing the rulers of Europe characterized these movements.

D-8. New concepts in the 17th century stressing the importance of the individual led to the formation of reform movements which brought about major changes in many countries. The overthrow of Charles I of England, for example, was affected primarily by religious groups which had been forced to meet clandestinely and which turned to political and military action in order to achieve a wide range of political as well as religious reforms. Later, clandestine groups favoring the restoration of the monarchy obtained funds from persons friendly to Crown Prince Charles and infiltrated offices of the English Government. In 1660, after years of civil war and interregnum, they brought Charles to the throne.

D-9. In the 18th century are two outstanding examples of underground movements that were well-organized and active on a wide variety of fronts. In America, the Secret Committee for Correspondence and affiliated groups evolved into an inter-colonial movement engaging in large-scale propaganda and other psychological warfare campaigns against the British colonial authorities. They trained militia groups, established friendly contacts abroad, and eventually played a leading role in gaining independence for the 13 colonies. In France, the Revolution of 1789 had its genesis in underground groups; once in control, the Jacobins organized networks of sympathizers in other countries dedicated to subversion and revolution. The subsequent overthrow of the radical Jacobins was followed by the establishment of the 5-man

Directory, which was soon in difficulties both at home and abroad. Disorder was rampant, the financial situation was desperate, discontent was everywhere, and the Republican armies suffered reverses. The government, not knowing which way to turn, swayed one day to the Jacobins, the next to the moderates. Under these circumstances, General Napoleon Bonaparte, with the aid of several political leaders, was able to come to power through a coup-d'état.

D-10. Napoleon's conquests into the early 19th century gave rise to a series of resistance movements throughout Europe and Latin America. In such notable cases as Spain, Mexico, and Venezuela, armed guerrilla uprisings occurred in the name of restoring or gaining national independence. The "Spanish Ulcer," as Napoleon termed the war in Spain, was one of the principal factors leading to the defeat of the French Emperor. France invaded Spain, defeated the regular army, and captured the main cities and towns. However, it was not able to control the people living in the rugged Spanish countryside. They organized into small irregular bands and made surprise attacks upon units of the French Army. They were estimated to have taken a total of 100 soldiers a day. This guerrilla war diverted French military resources badly needed for operations elsewhere. With the aid of English forces in Portugal and Spain, the Spanish guerrillas were able to drive the French from Spain and assist Wellington's forces in defeating Napoleon. "Guerrilla" is a Spanish word meaning "little war," or "body of partisans," and it was here that the term guerrilla was first used.

D-11. After the Congress of Vienna in 1815 and with the subsequent spread of nationalist aspirations, underground resistance and revolutionary movements became particularly important in effecting political change. Especially noteworthy are the Carbonari and Young Italy movements on the Italian peninsula, which were led by such men as Mazzini and Garibaldi These men sought not only independence for the Italian states under foreign rule but also the ultimate unification of Italy into one nation. The Carbonari began during the Napoleonic era as an open resistance movement with cells throughout most of Italy. In 1818, after an abortive insurrection, they were driven underground. Though operating clandestinely, the movement managed to enroll nobles, army officers, small landlords, government officials, peasants, and priests. It was not until 1861 that the goals of the movement were achieved.

D-12. Although unsuccessful for many years, the Irish nationalist movement attracted considerable attention throughout the 19th century by its constant use of political warfare and terrorist techniques. The Irish movement for independence was certainly one of the most prolonged of its kind, spanning nearly four centuries.

D-13. Underground movements throughout the Russian Empire during the late 19th century, many of which employed extensive terrorism, were almost as influential as Marxism itself in shaping Lenin's concepts of party organization and tactics. This was particularly true with regard to the use of terror for obtaining popular compliance with party demands. The Bolshevik revolution was brought about by well-organized underground activity led by Lenin's vision of a professional vanguard party.

D-14. World War I was precipitated in the early 20th century by the assassination of the Austrian Archduke Francis Ferdinand by a Bosnian student working with the Black Hand Society. This clandestine group, organized in Serbia, financed by the Russian-led Pan-Slav movement, and including among its members several high Serbian officials, was dedicated to liberating Slavs in the Austro-Hungarian Empire and organizing them into a Greater Serbia. The Great War in Arabia also witnessed one of the great exemplars of UW in T.E. Lawrence.

D-15. From this brief historical survey, it is clear that clandestine organizations conducting UW activities are not the product of a particular political or religious ideology; cultural, ethnic, national, or geographical grouping of persons; structure or form of government; segment of society or social class; or stage of a society's economic or technological development. Also it is clear that subversion, sabotage, and political resistance work of undergrounds and the guerrilla warfare-type tactics of raid and ambush are neither new nor unique to the contemporary world scene, although such an impression is easily created by the pressure of current problems. The activities associated with UW have appeared in political life throughout recorded history.

SELECT EXAMPLES FROM THE MODERN ERA OF INSURGENT, RESISTANCE, OR REVOLUTIONARY WARFARE

D-16. The best-known era of UW activities is the "classic era" during WWII. There are many sources of information on the exploits of the British Special Operations Executive, the American Office of Strategic Services, the French Resistance, Soviet Partisans throughout Eastern Europe, the American and Philippine Anti-Japanese Resistance, the Chinese Anti-Japanese Resistance, and Office of Strategic Services Detachment 101 in Burma.

D-17. Following the WWII period, insurgency and revolutionary warfare theory and practice flourished during the Cold War and the anti-colonialist uprisings around the globe. Some of the most famous, influential, and enduring insurgent and resistance concepts which have influenced thinking on UW came from this period. Finally, a few of the current theorists of insurgency, resistance, revolutionary warfare, and jihad are noted. The user of this publication is encouraged to research these topics and the following selected list of significant personalities and events:

- *Ayman Al-Zawahiri* (Egyptian): Physician, member of Egyptian Muslim Brotherhood, founder of the Egyptian Islamic Jihad, ideological and current political leader of al-Qaeda. Most important work is Knights under the Prophet's Banner. Key ideas include democracy must be destroyed because it says man rather Allah is sovereign, that jihad must be global, all Muslims everywhere must be held responsible for conducting jihad wherever possible, and that patience combined with mass killings through suicide bombings is required.
- *Abdullah Azzam* (Palestinian): Credentialed and authoritative Islamic scholar and theologian, teacher and mentor of Osama bin Laden, and most important figure in the creation of al-Qaeda. Most important work is The Defense of Muslim Lands. His most important contribution is he made the jihadist canon an internationally recognized concept; he was able to inspire and recruit fighters from across the Muslim world for Afghanistan. Azzam says jihad is obligatory and that no individual needs another's permission to wage it.
- *The Coming Insurrection* (France 2007): Anarchist work by a French so-called "Invisible Committee" which proclaims the coming collapse of global capitalist society. Political organization will devolve around clandestine underground identity groups which will both seize local power and disrupt larger-scale mechanisms of industrial civilization.
- *Dau Tranh* (Vietnamese): Vietnamese variant of People's War. Central idea is that everything conceivable is an instrument of comprehensive struggle; there are no "noncombatants," everything is a political act, and the total mobilization of a society's resources and psychic energies.
- *Ernest "Che" Guevara* (Argentinean): Marxist theorist, revolutionary and guerrilla leader, physician and author. Successful leader of the guerrilla campaign under Castro which led to communist victory in Cuba. Unsuccessful in exporting his techniques to Congo, Bolivia, and elsewhere. Author of *Guerrilla Warfare: A Method*. Major variant of revolutionary warfare was his discredited (although popular) "foco" theory which maintained that the armed struggle of rural guerrilla warfare vanguard elements could spark larger revolution.
- *Ho Chi Minh* (Vietnamese): Marxist revolutionary, leader of the Viet Minh, and victor of Indochinese wars against France, the United States, and South Vietnam. President and founder of the Democratic Republic of Vietnam. Central figure of Vietnamese communism.
- *S. K. Malik* (Pakistani): As a serving Pakistani general, wrote *The Quranic Concept of Power*, a book that was endorsed and introduced by religious authority in Pakistan and the then President and Commander-in-Chief of Pakistan's armed forces. Serves as a strategic vision for jihad claiming that the center of gravity in the struggle is the enemy's heart, mind, and soul; that the object is to defeat that heart, mind, and soul; and that the primary means to that end is terrorism. Malik describes and advocates "holy war."
- *Mao Tse-tung* (Chinese): Communist revolutionary, political philosopher, military strategist, founder and leader of the People's Republic of China from the 1920s to 1970s. Successful anti-Japanese resistance leader and Chinese Civil War victor. Notable for Marxist variant which articulated revolutionary power of Chinese mass populace rather than the orthodox view of

industrial proletariat of Marx or the revolutionary vanguard of Lenin. His Selected Military Writings articulates arguably the most influential theory of insurgency, resistance, and revolutionary warfare ever advocated—protracted people's war. Mao emphasized the inseparability and primacy of political work in revolutionary warfare, and identified three stages: the strategic defensive, the strategic stalemate, and the strategic offensive—otherwise known as the "latent and incipient," "guerrilla warfare," and "war of movement" phases respectively.

- *Carlos Marighella* (Brazilian): Marxist revolutionary and writer. Most famous work was *Minimanual of the Urban Guerrilla*, which articulated his theories of urban guerrilla warfare in order to disrupt and overthrow an authoritarian regime.
- *Sayyed Qutb* (Egyptian): Islamist theorist, educator, leading member of the Egyptian Muslim Brotherhood, and influential ideologue for jihad. Key work is *Milestones*, which is a fundamental roadmap for subsequent jihadist thinking and action. His essential thesis is that the world is heretical, it is to be forcibly cleansed by Islam, and Islam is not to be understood as just a religion but as a revolutionary party. Furthermore, Islam is the last message given and is ordained for mankind. Anyone who hinders the spread of Islam must be fought until he is killed or declares his submission. Ironically, Qutb's theories were influenced by Lenin's concept of the revolutionary vanguard.
- *STUXNET*: Computer worm surreptitiously released into Iranian nuclear program networks by unknown actors to covertly sabotage the industrial control systems for gas centrifuges. This type of sophisticated emerging technology has enormous implications for future insurgency.
- *Josep Broz Tito* (Yugoslavian): Communist Commander-in-Chief of the Yugoslavian National Liberation Military Forces from 1941 to 1945. Successful Anti-Axis partisan leader.
- *Vo Nguyen Giap* (Vietnamese): Principal Commander of the Viet Minh and the People's Army of Vietnam. Victor of the Indochinese wars against France, the United States, and South Vietnam. Most famous work is *People's Army, People's War*.

This page intentionally left blank.

Appendix E
Revolutionary and Insurgent Warfare Analysis

We should also remember that modern military technology, the information "battlespace," no matter how swift and powerful, has yet to replace the timeless necessity to understand the generic conflicts that have made guerrilla warfare and insurrection an enduring problem of history, politics, and culture.

John Tierney, Chasing Ghosts

This appendix provides description and analysis of numerous modern insurgencies and revolutions demonstrating current or recent UW potential. Each example is unclassified to provide the widest-possible dissemination and discussion. The analysis of each example is serious but subjective. Every opportunity to employ UW will be different. The appendix is meant to present the variety of insurgencies and revolutions to aid Soldiers' ability to think—not to provide a final judgment on what to think.

CRUCIAL AND CONTRIBUTING ANALYSIS FACTORS

E-1. The desire to change the style of a nation's government has been a cause for revolution dating back millennia and encompasses the more celebrated revolutions in history, such as the American revolt and the French overthrow of their monarchy in the 18th century, the European social revolutions in the 19th century, and the communist and anti-imperialist revolutions of the 20th century. The desire to overthrow and replace an entire system of government requires two crucial factors: an existing type of government that does not meet the needs of the population (or powerful segments thereof), and an ideologically mature alternative government to be proposed and rallied around.

E-2. The first factor requires that the populace (or important segments such as political elites, the merchant class, blue-collar workers, academia, and so on) be unhappy in some fashion with their present lot in society. Their displeasure may be rooted in socioeconomic conditions, such as an economic depression, lack of political influence or representation, and large disparities between upper and lower economic classes. Or, it may be displeasure with the actions of the government, such as excessive taxation, harsh security measures (such as curfews and martial law), or even dramatic changes in policy, such as the introduction of modern economic measures or restriction of goods and services once provided. This motivated segment of the population becomes the recruitment base for the fomenting revolution. Support networks often operate purely within those social groups throughout the revolution, rarely expanding into other segments unless the revolution truly becomes a populist wave. Many examples exist where this group is a well-educated segment of the population, where their relative disillusionment or disenfranchisement with the government is the motivating factor.

E-3. The second factor for a revolution to modify the government type is an ideologically different government system that has either worked in other nations or has reached a level of theoretical maturity that it can convince a large group of people that it would be successful. In many cases, the ideological form of government is a well-established socialist construct, usually traceable to some form of Marxism or its main variants which assert that the population would benefit from a less-capitalistic economic system to one that is more communal and centrally governed. Such an economic class-based viewpoint provides a useful baseline for analysis, particularly as many of the most famous and successful revolutionary efforts and figures of the 20th century were Marxist revolutionaries.

Appendix E

E-4. However, not all revolutions are Marxist based. For example, Iran and Burma represent cases where the unifying motivation was or is opposition to dictatorial regimes. The supremacy of the Iranian ayatollahs was also a foretaste of the ongoing revolutions throughout southwest and southern Asia based on radical Islamic fundamentalism. Many of the revolutions of the past century, such as in Vietnam or 1980s Afghanistan, were essentially nationalist or patriotic efforts to expel foreign powers. Some revolutions, such as those in Ireland or Sri Lanka, were based on identity issues or ethnic grievances. Finally, some revolutions such as Solidarity in Poland or the Orange Revolution in Ukraine sought reforms and liberalizations rather than outright government replacement. More detailed discussion of analyzing revolutionary and insurgent warfare is contained in *Casebook on Insurgency and Revolutionary Warfare, Volume 2: 1962–2009*.

MOTIVATIONS

E-5. A myriad of contributing factors combine to form each distinct case of revolutionary warfare or insurgency, and it is essential for those conducting UW to understand the characteristics of the problem as it is, not as simply a representative sample. Most often, any given example of revolutionary warfare or insurgency will exhibit a combination of motivations at least as numerous as the various rebellious factions involved. However, some structure for analysis and discussion of similarities and differences is required. This publication categorizes revolutionary warfare and insurgency by the most evident or persuasive cause for which the revolt occurred. The five primary motivations are—

- A desire to greatly modify the type of government. Representative examples include the New People's Army (NPA) in the Philippines; the FARC; Sendero Luminoso (Shining Path) in Peru; the 1979 Iranian Revolution; Farabundo Marti Frente Para la Liberacion Nacional (FMLN) in El Salvador; and the Karen National Liberation Army (KNLA) in Burma.
- Revolutions where identity or ethnic issues are primary. Representative examples include the Liberation Tigers of Tamil Eelam (LTTE) in Sri Lanka; the Palestine Liberation Organization (PLO); the Hutu-Tutsi genocides in Rwanda; the Kosovo Liberation Army (KLA); and the Provisional Irish Republican Army (PIRA).
- A desire to drive a foreign power out of their area. Representative examples include the Afghan Mujahidin in the 1980s; the Vietcong in South Vietnam; the Chechen Revolution; Hezbollah in Lebanon; and the Hizbul Mujahedeen in Kashmir.
- Revolutions based upon religious fundamentalism. Representative examples include the Egyptian Islamic Jihad (EIJ); the Taliban in the 1990s; and al-Qaeda.
- Revolutions primarily seeking modernization or reform rather than overthrow. Representative examples include the Movement for the Emancipation of the Niger Delta (MEND) in Nigeria; the Revolutionary United Front (RUF) in Sierra Leone; the Orange Revolution of Ukraine; and Solidarity in Poland.

OPERATIONAL ENVIRONMENT

E-6. Every revolution is a reaction to a particular governing and political environment. The Soldier conducting UW must explore and understand the operational environment unique to the operational area being considered. Only by understanding the operational environment can the appropriate UW objectives and the required target audiences for influence be determined, and the multiple lines of effort be developed to meet campaign objectives.

E-7. Relevant information on the operational environment consists of the eight operational variables of PMESII-PT. Those variables could influence or determine how the revolution or insurgency originated, what its limitations and constraints are, and how the context might be manipulated to achieve desired results. Understanding of the pertinent regional and local history is crucially important. Historical information can explain causes and effects that are important for highlighting motivational issues such as long-lasting grievances, and often provide the central insurgent narrative. Understanding the psychological effects of these narratives requires a sophisticated appreciation of the operational environment and promises nonlethal methods of influence crucial to the long-term conduct of UW.

OBJECTIVES AND GOALS

E-8. The strategic objective of revolutions and insurgencies is to gain power, whether the movement seeks a total overthrow of the existing government and replacement with something entirely different, as is the case of the communist and Islamic fundamentalist movements, or more limited objectives, such as reform of the existing system for economic, ethnic, or political reasons. Specific objectives vary based on the context unique to each case and may change over time based on changes in that context or the changing fortunes of a given movement.

E-9. Many of the communist movements used similar operative strategies and sought similar intermediate objectives. For example, the NPA, FARC, Sendero Luminoso, Viet Cong, and the FMLN all sought to organize masses of rural peasants to build their power base while slowly constricting the cities, degrading the national infrastructure, and undermining the capabilities and authorities of the state. Nevertheless, each movement had its own unique characteristics. One important unifying goal of the NPA was to depose the Marcos dictatorship en route to a communist system. Once Marcos finally stepped down and reforms were enacted, however, the NPA found much of its appeal diminished. The FARC was only partly inspired by Marxism; it was in varying degrees inspired by Bolivarian nationalism and outright criminality based on the cocaine trade. Sendero's focus on "pure native peasant communism" and longstanding self-imposed international isolation both constituted built-in limits to movement expansion. Moreover, the NPA, Viet Cong, and FMLN deliberately used a mixture of rural and urban operations that varied from the classic Chinese protracted war model.

E-10. Many revolutionary movements determined that separating the existing state's external support was a critical goal. The Viet Cong, FMLN, opponents of the Shah, EIJ, and al-Qaeda all made the degrading and forcing out of U.S. support a key priority. Many movements' primary goal was the expulsion of "foreign occupiers." The Taliban sought the removal of the Soviets in Afghanistan. The PIRA sought the end of British Army occupation in Northern Ireland as it represented an obstacle to Irish unification. The PLO regards Israeli occupation of the West Bank as illegal occupation, and attempts to force "right of return," "2-state solutions," and proposed withdrawal to pre-1967 borders all seek the objective of weakening Israeli physical security.

E-11. Several modern-day revolutionary movements, such as the LTTE, Karen National Union, and KLA, sought outright independence or significant autonomy as their end state objective. Motivations and short-term objectives changed with conditions, however. In their initial push for Tamil autonomy, the LTTE became the unifying opposition to the Sinhalese majority only after accomplishing the near-term goal of eliminating all other Tamil rivals. After India imposed a peacekeeping force in northern Sri Lanka to constrain years of suicide bombings, assassinations, and widespread carnage, the LTTE shifted its operational objective to attacking the Indians—in temporary alliance with Sinhalese nationalists—until the Indians left. Once that objective had been achieved, however, the LTTE resumed its attacks against the government.

E-12. One supporting objective of many insurgent groups was the calculated exploitation of IO and international sustainment activities. The LTTE were adept at international fundraising, particularly in Europe and Canada. The NPA and PIRA drew support from the ethnic Filipino and Irish diasporas, particularly in America. The KLA and PLO placed an emphasis on careful coordination of propaganda explanations of their operations for the purpose of creating international sympathy. The PLO, Hezbollah, and al-Qaeda have all been skilled in the goal of drawing support from the larger Muslim world and Europe by careful manipulation of messages.

E-13. Finally, movement goals and objectives should not be assumed to necessarily reflect western democratic values; the revolutionary motivation of "popular emancipation" assumed by western countries may not be accurate. Some revolutionary and insurgent movements reflect base or evil motivations. Marxist insurgencies claim to be about "liberation," but usually resort to terror and murder to meet the goal of state dictatorship. Radical Islamic movements seek the objective of destroying western state sovereignty and imposing sharia law at the expense of individual liberty. The goal of the Rwandan massacre was deliberate Tutsi genocide at the hands of the Hutu. The "just peoples' movement" goals of the FARC,

MEND, and RUF were convenient excuses for the more venal objectives of controlling cocaine, oil, and diamond wealth respectively.

LEADERSHIP AND ORGANIZATIONAL STRUCTURE

E-14. There are a few essential leadership and organization characteristics of a revolutionary movement or insurgency that a Soldier planning for UW should determine; each of these is discussed in the following paragraphs. Ultimately, combinations of actual characteristics specific to a given AO are more important and likely than a fixed theoretical organizational template.

HOW THE ORGANIZATION STRUCTURE REFLECTS THE UNDERLYING MOVEMENT PHILOSOPHY, IDEOLOGY, OR STRATEGIC APPROACH

E-15. The structure of revolutionary organizations and insurgencies typically reflect the movement's underlying philosophy, ideology, or strategic approach. Mass-based Maoist-inspired insurgencies like the NPA, Sendero Luminoso, FMLN, or the Viet Cong will attempt mass organization characterized by large, labor-intensive structures and presence diffused down even to low-level villages. Such organizations seek to win by demographic constriction of the government and require large political cadres to preach and uphold a political orthodoxy amongst a multitude. The organizational structure of Hezbollah provides an interesting variant of the mass-oriented approach. Hezbollah's structure includes a network for funding sources and a provision of social services. It has morphed the organization from a resistance movement into a stateless government that leverages charity to meet the needs of the neglected, oppressed, and marginalized Lebanese Shia population to maintain popular support. This structure supports the Islamist narrative of a pervasive force that is penetrating all aspects of society. By contrast, an elite vanguard-oriented movement like the Bolsheviks will have a relatively small organization and seek to achieve a quick "surgical" seizure of power. Security and proselytizing are less problematic as such organizations are generally limited to only those totally and irreversibly committed to the cause. Some movements based on religious or ethnic affiliation like the Chechens, Mujahedeen, al-Qaeda, LTTE, or KLA assume philosophical adherence and can more likely be organized along traditional representative structures of control. Finally, some movements like the 1979 revolution or PLO are umbrella organizations that seek to collectively represent ideologically diverse components unified by a common enemy.

HOW DEMOGRAPHICS AND POLITICAL REALITIES SHAPE THE AREA OF OPERATIONS

E-16. Revolutionary and insurgent structures are determined—at least in part—by the political, social, and security realities of the AO. Movements built on religious ethnic or religious grievances, such as the LTTE, Chechens, PIRA, or EIJ, are almost certain to be organized along ethnic and religious lines. Predominantly mass-based movements, such as the FARC or Viet Cong, are typically centered in rural areas and difficult hinterlands at the extremes of state coercive power. Conversely, predominantly urban-based movements, such as Solidarity or the PIRA, rely on underground and cellular structures when confronted by effective counterinsurgent power. Some movements, such as Hezbollah or the PLO, deliberately balance overt political activism and community organizing with the application of guerrilla or terrorist violence.

THE NATURE OF THE SENIOR LEADERSHIP (PROFESSIONAL AND BUREAUCRATIC OR ESSENTIALLY CHARISMATIC)

E-17. All revolutionary or insurgent movements require some form of leadership, but the character of that leadership will vary. For mass-based efforts, such as the NPA or Viet Cong, the emphasis is on efficient organizing and fastidious attention to regimented political orthodoxy; adherence to ideological correctness is paramount. Islamic radical-inspired movements, such as the EIJ, the Taliban, Hizbul Mujahedeen, or al-Qaeda, have a similar emphasis on doctrinal correctness. In both such cases, followers are inspired as much by their allegiance to ideas—or their fear of discipline—than by allegiance to a particular personality. By contrast, some movements, such as Sendero Luminoso, operate as virtual personality cults dominated by key personalities like Abimael Guzman. In addition, some widespread movements like the

1979 Iranian revolution, Solidarity, or al-Qaeda become symbolically associated over time with internationally recognized figures as Khomeini, Walesa, or Bin Laden.

THE ORGANIZATION (RELATIVELY SIMPLE OR COMPLEX)

E-18. Some movements, such as the Chechens, Afghan Mujahedeen, or the Taliban, benefit from straightforward motivations (such as religious fervor and desire to expel invaders) and in-place social structures (such as local clan allegiances) that simplify the organizational task of insurgency. By contrast, some movements, such as the 1979 Iranian revolution or the PLO, are complex umbrella coalitions of diverse and fractious components. In the case of Iran, component parties unified in their opposition to the Shah were as diverse as Marxists, monarchists, anti-Shah secularists, and Islamic fundamentalists. Likewise, the PLO represented factions who advocated or rejected violence, Islamic fanatics, anti-Semites, and secular Palestinian nationalists. Some movements were also relatively simple in strategic approach, such as the Sendero who relied on savage orthodoxy and reclusive dismissal of outside assistance, or the Hutu Interahamwe who were unified by the simple intent to commit genocide everywhere in Rwanda. By contrast, many other movements like the FMLN, LTTE, Viet Cong or KLA combined violence with nonviolence, a mixture of rural and urban operations, and employed strategies of internal and international strategic communications and material support.

LEADERSHIP AUTHORITY AND ORGANIZATIONAL CONTROL (CENTRALIZED OR DECENTRALIZED)

E-19. Almost all revolutionary or insurgent movements have some degree of centralized strategic direction and planning. In the case of the mass-based Marxist movements, such as the NPA or Viet Cong, centralized leadership provides strategic direction and this guidance is meant to flow down unmodified through a hierarchical leadership and political structure embedded at every echelon. Variations to this model exist. On one extreme, the FARC managed to maintain considerable overall strategic coherence based on their approach of armed struggle while being ideologically fragmented. This lack of ideological rigidity benefited recruitment. In fact, FARC maintained a deliberate distance from the Colombian Communist Party and allowed it to lead in political activities and ideological purity. By contrast, the Sendero Luminoso enforced a savage and fanatical ideological rigidity which imposed self-inflicted limits to recruitment, external support, and organizational growth. The al-Qaeda provides a different but important example of centralized ideology but decentralized execution. The motivating ideological core and strategic direction of al-Qaeda-affiliated operatives was established by a small percentage of imams and theorists, and certain operations were funded and organized from this central strategic leadership. However, al-Qaeda operates in many respects as a leaderless organization, relatively unconnected, and secure cells conducting decentralized initiatives in operations within the context of an agreed-upon strategic goal.

NOTEWORTHY ADVANTAGES OR DISADVANTAGES RESULTING FROM THE STYLE OF LEADERSHIP OR CHOICE OF ORGANIZATION

E-20. The hierarchical mass-based systems can theoretically provide a huge demographic advantage when actions are successfully coordinated. However, the practical political, communication, and logistical challenges of such mass organizations are time-consuming and ponderous. As seen in the NPA, Viet Cong, and all other Marxist organizations, widespread and continuous conformity can only be achieved and maintained through systematic and ongoing coercion. Sendero and the NPA are both examples of fanaticism and murderous purges intended to enforce party purity actually retarded their organizational growth. The 1979 Iranian Revolution and the PLO both represent coalitions comprised of diverse and opposing interest groups that through the signature leadership of Khomeini and Arafat achieved more than the individual factions could have separately. As already noted, al-Qaeda's decentralized execution structure enhanced its security and had the bonus of magnifying the perception of its global reach.

Appendix E

COMMUNICATIONS

E-21. Communications is divided into two main considerations: means of delivery and content. Revolutionary and insurgent communication delivery techniques run the gamut from ancient face-to-face meetings based on long-established social constructs to state-of-the-art telecommunications, high-technology equipment, professionally developed advertising, and global media management. Content themes and narratives are particular to each revolution or insurgency and vary widely in sophistication.

COMMUNICATION DELIVERY TECHNIQUES

E-22. Common rudimentary communication methods include face-to-face meetings, couriers, mass meetings, pamphleteering, newsletters, newspapers, tactical radio communications (often from radios stolen from the opposing government armed forces), simple radio broadcasts, and in recent years common and inexpensive unsecure cell phones and the internet. Most recent revolutionary movements or insurgencies have used some or most of these techniques. There are some movements, however, which illustrate peculiar or exceptional examples.

E-23. The fanatic attention to security in the Sendero Luminoso severely limited its communications. Sendero was organized into operational cells of 10 to 12 members. Only the cell leader was authorized to communicate with other Sendero members outside of the cell, and then only through aliases. While this limited organizational knowledge and mitigated risk from compromise, it also made communication a ponderous process. Moreover, since all contact was funneled through one key individual, loss of that individual could render the entire cell inoperative.

E-24. Many Muslim movements enjoyed communication channels already existent through religious observance. Islam is a proselytizing religion that demands continuous daily ritual attention and in which the guiding role of imams and the centralizing focal point of the mosque is woven into society. This ritual observance and organizational routine provides a built-in communication channel. Islamic movements' communication methods should not be dismissed as only religious or only low-technology, however. The Iranian Revolution also benefited highly from the societal practice of going to the bazaar—a perfect locus for exchanging goods, services, and messages supportive of revolution.

E-25. Some Islamic movements are also very adept at modern communications. Hezbollah's communication network constitutes the largest network of any political party in the region consisting of a satellite channel with over 10 million viewers, 4 licensed radio stations, and 5 licensed newspapers; all of which generate significant self-supporting income through advertising. Hezbollah's graphics and editing are state-of-the-art. It employs foreign correspondents throughout their region, Europe, and North America. Hezbollah's use of cell and satellite phone technology is widespread despite its vulnerability. Hezbollah has gradually replaced vulnerable copper landlines with fiber-optic cable. Hezbollah has even developed electronic warfare methods capable of interfering with some Israeli communications while protecting its own. Finally, Hezbollah exploits cyberspace. Its hackers routinely hijack and exploit other Web sites, and foreign journalists seek out its web content based on a reputation for access and credibility. Hezbollah's Secretary General Nasrallah underscored the centrality of the group's communication network in 2008 when he declared it to be the group's single most important weapon, the disruption of which would constitute an act of war.

E-26. One counterintuitive example of communication delivery is worth noting as a modern UW venue that illustrates classic UW principles. The Ukraine state authorities controlled practically all media in 2004, and tightened its repression as opposition strength grew. In addition, only two to four percent of Ukrainians had access to the Internet. Both facts would suggest the unfeasibility of revolutionary communications. However, most of the few Internet users were concentrated in the capital, Kiev. These Internet users retransmitted useful information via word-of-mouth, cell phone, text messaging, and other more common techniques. These "retransmitters" thus functioned as force multipliers for the revolution. Even more importantly, by repressing modern technology in and out of the capital, opposition leader Yushchenko was forced to take his message to the countryside in a long series of face-to-face campaigning and organizing. Ultimately, the popular organizing and motivation resulting from this old-fashioned campaigning would prove crucial to organizing the final positive outcome of the Orange Revolution.

COMMUNICATION CONTENT

E-27. The Iranian Revolution was largely based on opposition to the secularizing force of the Shah's policies. Since opposition was organized predominantly through religion, counterrevolutionary or counterinsurgent activity by the state could be easily interpreted as anti-Islamic repression, thus reinforcing the insurgents' main theme. To further exploit this dilemma for the state, Khomeini's forces shrewdly scheduled demonstrations on holy days or days scheduled for martyrs' observances.

E-28. A classic trick of political persuasion is to give a slanted version of events that are carefully selected and shaped to encourage sympathy for the revolutionary movement or insurgency, while putting the government—and rival contending groups when necessary—in an unsympathetic negative light. Marxist revolutions of all varieties everywhere (NPA, Viet Cong, FMLN, FARC, and Sendero Luminoso) all emphasize a theory of class warfare and repression of large poor masses by small rich elites. Radical Islamic movements (EIJ, Taliban, Hezbollah, Hizbul Mujahedeen, and the Chechens) all emphasize achieving or returning to a state of Muslim authority and autonomy versus secularist and apostate governments. Ethnicity and secessionist-inspired revolutions (LTTE, KLA, PIRA, Rwanda, PLO, and the Chechens again) seek to highlight a claim to autonomy and gain sympathy through a claim that the state power of their oppressors is historically unjust and inhumanely disproportionate. Some examples stand out.

E-29. Some movements have been notably successful in gaining external support or even international sympathy. Combining a rapid and comprehensive method of information dissemination to expatriate Tamil communities with ruthless elimination of domestic rivals, the LTTE was able to achieve the leadership of Tamil resistance. Through a carefully constructed narrative of historical rights to their land and a readily accepted characterization of ongoing Serbian brutality, the KLA was effective in gaining international support. The PLO has been similarly effective in gaining international support against Israel through a narrative claiming illegal occupation of Palestinian territory. Solidarity was likewise effective in gaining international sympathy against a long context of Polish and Soviet communist oppression, especially by gaining the support of the Polish-born Catholic Pope.

E-30. Two of the most effective and sophisticated Islamic examples of powerfully persuasive content are Hezbollah and al-Qaeda. Combined with sophisticated and rapid-delivery architecture, Hezbollah has been notably adept at shaping tactical narratives that enhance their strategic claims against Israeli "hegemony" and Western "imperialism." Hezbollah routinely edits conflict footage to its advantage by always portraying Israeli actions as an aggressor and in the worst possible interpretation. Conversely, Hezbollah crafts footage to highlight the theme of heroic resistance by "oppressed" Arabs, most often by virtually unarmed youth in the face of Israeli tanks and warplanes. Hezbollah routinely stages protests and fakes casualties for propaganda purposes. All of these tactical efforts combine to gain noticeable strategic support for the movement. Moreover, Hezbollah maintains a steady campaign to undermine Israeli domestic morale through taunting emphasis on current and future Israeli casualties.

E-31. Most recently, al-Qaeda has been effective in inspiring Muslims worldwide through an overarching and constantly repeated strategic narrative. The main tenets of this narrative are that countries in the region, such as Saudi Arabia, Egypt, and Pakistan, have apostate regimes that work in concert with the United States and other western countries to repress Muslims; that such "Crusaders" have worked to prevent true believers from establishing an Islamic state; that Christendom along with world Jewry is seeking to destroy Islam; that the United States is the cause of most such evil and deserves attacks such as those inflicted on September 2001; that the economy of the United States is the U.S. center of gravity and must be attacked; and it is the solemn duty of all Muslims everywhere to kill westerners—especially Americans—anywhere and anytime they can. This narrative continues to be persuasive to millions of al-Qaeda sympathizers worldwide.

E-32. Finally, it is worth observing how revolutionary or insurgent narratives—and their deliberate application and repetition—can serve to prepare movements of many kinds to commit violence, up to and including the genocides of the communists, Nazis, or Khmer Rouge. In Rwanda, long-festering ethnic animosities between Hutus and Tutsis were deliberately and methodically shaped into an overheated atmosphere for genocide. Latent Hutu-Tutsi animosity was made critical by imperialist Belgian efforts to divide them along exaggerated race lines, including a myth that the Tutsi were a "Hamitic race" that originated elsewhere and which helped justify Belgian support for Tutsi rule and Hutu subordination.

E-33. Hutu academics wrote a Hutu manifesto in 1957 that codified this myth but interpreted it as why Tutsi rule and presence was illegitimate. From then on, Hutus were taught to acknowledge their majority status and prepare for future ethnic violence. In 1990, a "Hutu Ten Commandments" promoted ethnic separation because Tutsis were "traitors." A comprehensive agitation campaign by Hutu extremists in and supporting the government used oral announcements, leaflets, and all the main radio broadcasts to advocate Tutsi extermination. When Hutu gangs murdered thousands of Tutsis in various incidents, the violence was blamed on the "savagery" of the Tutsis themselves. In the months prior to the 1994 genocide, radios broadcast false accusations and hate-filled diatribes against "alien" and "evil" Tutsis and an approaching "final battle." As the genocide began, government-organized Hutu murder gangs were joined by civilians who were encouraged 24-hours-a-day to join the gangs as "real men" and "defenders" to murder Tutsis wherever they could be found. Radios broadcast Tutsi license plate numbers, addresses, directions to find Tutsi homes, and instructions on how to construct roadblocks and checkpoints for an anti-Tutsi dragnet. Such deliberate inflammatory content contributed to an estimated 800,000 murders within 100 days.

METHODS OF ACTION AND VIOLENCE

E-34. Compared to the strength of state security forces, insurgencies and revolutionary movements almost always suffer a prohibition of overt oppositional activities and an inferiority of force. Not surprisingly, the standard methods of organization are typically discrete and underground, and the standard methods of armed violence are those of low-level criminality, terrorism, and guerrilla warfare. Typical methods of nonviolent action include subversion in many forms, including political organization and manipulation, political agitation, propagandizing, and fund raising. Typical methods of violence that are shared by most insurgent and revolutionary movements include kidnapping, extortion, targeted assassination, torture and brutality for terrorist psychological and political effect, bombings, and the classic guerrilla warfare techniques of raid and ambush.

E-35. Methods of action and violence are often determined by the operational environment. The NPA realized that maintaining an insurgent organization across the 7,100 Philippine Islands would be very difficult. The NPA shrewdly decided to turn an operational disadvantage into an advantage by decentralizing organizational control. Not only did this fit the practical and logistical realities of the archipelago, it had the intended effect of dispersing government security forces. Solidarity also had to constrain its methods due to the complete government monopoly of violence and the ever-present threat of massive Soviet intervention. Although unable to forcibly resist, Solidarity realized that its coercive power resided in the fact that its members comprised millions of industrial proletariat, a workforce upon which the Polish economy—and ultimately the Polish communist regime—could not function without. The Iranian revolution was another example where the state enjoyed a complete monopoly of violent coercive power. However, the mullahs were able to organize throngs of anti-secularist Muslim crowds whose numbers grew beyond the power of the state to contain.

E-36. Action and violence vary depending on the breadth and depth of popular displeasure with the targeted regime. Resistance to the Shah was widespread in Iranian society, and it took the combined resistance of all political groups to force the Shah to step down. Initially, Khomeini included non-Muslim but anti-regime entities in the short-term effort to depose the Shah, but then successfully seized control of the revolution and turned it into an Islamic regime. Ethnically and religiously centered revolutions, such as PIRA, KLA, or the Chechens, enjoy a natural unity of the movement's aims with a sympathetic and supportive kindred population. Mao said, "The guerrilla moves through the people as the fish moves through the sea." An aggrieved and defensive population from which the movement arises provides that "sea."

E-37. Violence is sometimes chaotic and largely indiscriminate. Violence in Sierra Leone was often little more than anarchic savagery by disparate greedy groups. The genocidal massacres of Rwanda were deliberate, planned, and coordinated, but the targeting involved was merely a blanket imperative to indiscriminately slaughter Tutsis wherever they could be found. Many bombings—such as those conducted by al-Qaeda or the PIRA—may have had deliberate symbolic targets in mind, but the actual detonations often indiscriminately killed and maimed any unfortunates who happened to wander into the kill zone.

E-38. By contrast, many movements—Hezbollah or most of the Marxist movements, for example—are especially deliberate in their targeting. Careful targeting of only security forces, specific assassination targets, and economic sabotage intended to weaken the state's political legitimacy are integrated actions subordinated to a larger strategic plan.

E-39. Hezbollah's strategy in defending against Israel's invasion in 2006 was a strategic success despite any significant tactical victories. Hezbollah intended its tactical actions to force Israel to commit more troops and resources to Southern Lebanon than the latter expected. Its deliberate targeting of reservist conscripts rather than professionals increased the war's perceived human cost among the Israeli public. Meanwhile, despite Israeli occupation of Southern Lebanon, Hezbollah continued to rocket Israeli communities from concealed locations. The cumulative effect of all of these tactical actions—each by itself of minor importance—was eventual Israeli withdrawal and defeat.

E-40. Perhaps the best example of deliberate and methodical targeting of action and violence was the Viet Cong within the North Vietnamese strategy of Dau Trahn. Dau Trahn coordinated every political, economic, and military action into a synchronized strategy. The Viet Cong played a critical role within this strategy at the village level. The Viet Cong eroded government power by slowly and methodically taking control of areas outside of constant government protection. Typically, the communists would infiltrate a village to identify who was hostile or sympathetic to their cause. Once opponents were identified, armed "agitprop" (agitation-propaganda) units would visit the village when government troops were not around. Torture killings of opponents—especially political leaders or authority figures like school teachers—would terrorize the village population into compliance and support of the movement. Once opponents had been removed, the village leadership would be covertly reorganized to support the revolution. From then on, supporters of the revolution and communist party would be promoted, while the remaining villagers would be monitored and made obedient by programmed indoctrination.

METHODS OF RECRUITMENT

E-41. There are many motivational factors of the operational environment which are typical of most revolutionary movements and insurgencies and which tend to contribute to recruitment. Common motivations that enable recruitment include—
- Antipathy to an individual ruling figure or regime.
- Government corruption and incompetence.
- Government and military human rights violations.
- Desire for security.
- Desire for revenge.
- Grievous socioeconomic disparities.
- Lack of legal jobs but a promise of illegal employment.
- Opportunity for adventure.
- Opportunity for status.
- Opportunity to find personal meaning in committing to a movement.
- Camaraderie.
- Opportunity to act on violent proclivities or engage in brigandage.
- Opportunity to protect the global commons, such as sea, air, space, and cyberspace.

E-42. There is also a variety of recruitment methods employed by various movements. In cases where resistance to foreign occupation occurs, such as the PLO, PIRA, Mujahedeen, Taliban, or the Chechens, the motivation is obvious and broadly unifying without need of excessive persuasion. The more brutal the occupation, and the more ethnically or religiously different the resistors are from the occupiers, the more this will be true. A similar commonality of broad-based motivation exists where ethnically, religiously, or socially distinct population groups seek some measure of independence or autonomy from a state

Appendix E

dominated by another distinct group. Examples of this include the LTTE, KNLA, KLA, and to some extent the Chechens and the Indian tribe-based Sendero Luminoso. In these cases, too, the motivations to join a movement are both obvious and widespread. In addition, reformist cases such as the Iranian revolution and Solidarity represent such widespread discontent that desire for change encompasses nearly every segment of society.

E-43. Almost all groups find the following methods useful in their recruitment, and the ones in parentheses specialize in using the approach listed:

- Face-to-face intimate appeals which develop individuals gradually and which exploit kin and friendship associations (EIJ, al-Qaeda).
- Broad-based appeals to ideological (the Marxist movements), ethnic (KNLA, KLA, Chechen, Hutus, LTTE, and others), or religious solidarity (Mujahedeen, Taliban, al-Qaeda, PIRA, Chechens).
- Coerced induction (some Marxist movements, especially Sendero Luminoso and the Viet Cong, the Hutu in Rwanda, and the RUF in Sierra Leone).
- Payment for participation (PLO, MEND, Taliban, FARC).
- Infiltration of legitimate organizations such as universities, student groups, and labor movements, or the creation of front groups to manipulate mass movements and as an opportunity to assess potential members (all of the Marxist movements, Hezbollah).
- Appeals to identity, meaning, community, paradise, and/or adventure (Sendero Luminoso, LTTE, al-Qaeda, Hutu).
- Appeals to bloodlust, pillage, and plunder (Hutus, RUF in Sierra Leone).

E-44. A few examples deserve special note. Some groups, such as the FMLN, eschewed forced recruitment because it was viewed as a security risk to have members participating solely through coercion. By contrast, Sendero Luminoso used forced coercion extensively. This was possible due to the absence of any government authority to prevent such coercion and was maintained by savage, rigid discipline. In addition, although many movements sought volunteers, Sendero was suspicious of them and would only take those which the movement had taken the initiative to select.

METHODS OF SUSTAINMENT

E-45. Methods of revolutionary and insurgent sustainment vary from the earliest and most tentative steps at local self-sufficiency to complex and sophisticated global financial networks. Some movements (the Iranian Revolution, Solidarity) are largely indistinguishable from the underlying sociopolitical structure and trends of political activity. Therefore, no additional funding or equipping is required. Other movements (LTTE, Mujahedeen, KLA) require large quantities of weaponry and supplies and rely predominantly on external sources of support. Some movements rely heavily on criminal activity for sustainment, whether large-scale narcotic involvement (FARC, Sendero, Taliban, Chechens, KNLA), theft and sale of natural resources (Chechens, RUF in Sierra Leone, MEND), or resort to simple brigandage (RUF, Hutus).

E-46. A majority of movements attempt to use some combination of the following basic sustainment methods: simple theft; raids on government facilities; battlefield recovery of supplies; revolutionary "taxes" on the populace, businesses, and/or landowners; extortion; forced labor; forced billeting; donations or tithes; embezzlement of legitimate enterprises; and deliberate domestic and international fund raising. A few movements present examples of broad and complex sustainment mechanisms that were notably effective:

- *Liberation Tigers of Tamil Eelam*: The LTTE conducted the standard local methods of sustainment, such as bank robberies, extortion, local taxation, and the smuggling of drugs and other contraband. However, the vast majority of the group's financing came from the large Tamil expatriate community in Western countries and the Indian province of Tamil Nadu. Expatriate Tamils included English-speaking professionals who had escaped the violence in Sri Lanka, plus refugees from such violence in the early 1980s. These communities thus had both the financial means to support and vivid memories of the urgent need. Expatriate support was sometimes voluntary and sometimes coerced and evolved in sophistication. The LTTE's reputation for ruthlessness contributed to "voluntary" donations, as did the fact that communities were organized and monitored—even by automated database and scheduled computer collection methods—for regular and continuous support. The LTTE developed a widespread network of suppliers and smugglers and maintained its own extensive maritime fleet of merchant ships, trawlers, fishing boats, motor launches, and even submarines. The LTTE's single biggest sustainment blow occurred when its support fell off sharply after being labeled a terrorist organization by dozens of countries in 2006.

- *Hezbollah*: This movement has been able to draw upon a wide range of financial sources, including funding from Iran, sympathetic donors in other Arab countries, and the group's own extensive licit and illicit business interests. Iranian resources have included money; hardware; training provided for military and resistance activities; the services of Iranian engineers, doctors, and other professionals; and financial services designed to help Hezbollah evade international sanctions. The movement also benefits from monies originating in private foundations and charitable organizations in Iran. Financing from Arab donors is primarily in the form of tithes from individuals, groups, small businesses, and banks in the Arab world and among the Shiite in the larger international community. Because Hezbollah is integrated into Lebanese society, it operates an extensive network of enterprises. The legitimate businesses include supermarkets, gas stations, department stores, restaurants, construction companies, travel agencies, as well as offshore companies, banks, and currency exchanges. In addition, several Lebanese financial institutions acted as intermediaries between Hezbollah and international mainstream banks. Hezbollah also profits from illicit activities, including smuggling drugs, diamonds, and other contraband, plus the receipt of laundered monies from the cocaine trade by the Lebanese expatriate community in South America.

- *al-Qaeda*: The al-Qaeda has developed sophisticated methods of sustainment that rely on both centuries-old patterns of indigenous commerce and modern methods of international finance. As a revolutionary or insurgent movement, al-Qaeda has been developing a system of training and sustainment camps since the Soviet invasion of Afghanistan. This clandestine network of support has exploited the traditional "hawala" patterns of commerce, has made use of the traditional Muslim requirement to tithe, has exploited mosques and religious institutions, and has created or infiltrated numerous international charities, banks, and front companies through which to funnel money. The al-Qaeda relies on an international system using modern methods of fundraising that is represented in over 100 countries. It has proven adept at security of these transactions by using a cellular structure, and using numerous personal, legal, and institutional cutouts. Moreover, far-flung operational cells were expected to be as self-sufficient as possible, thereby minimizing the occasional pattern of centrally provided organizational sustainment.

METHODS OF OBTAINING LEGITIMACY

E-47. Like other characteristics of insurgencies or revolutionary movements, what makes any particular movement legitimate is related to the underlying conditions and popular grievances of any particular situation. Nevertheless, the following generalities are most often true. Domestic legitimacy is a function of ability, veracity, and instrumentality meaning a movement must be effective, must be perceived to conduct activities congruous with its stated objectives, and must be useful to whoever bestows "legitimacy" on the movement. Therefore, movements that effectively do what they say on behalf of those they claim to represent will be judged legitimate. However, those movements that are incompetent or that become just another actor exploiting and repressing the people they claim to represent will be judged illegitimate. Internationally, legitimacy is a function of adhering to accepted international norms and laws, and having some reasonable basis for claiming a right to represent some demographic. In most cases, one can assume that the opponent state will deny the legitimacy of an insurgent or revolutionary group.

E-48. Some movements (Solidarity, Orange Revolution, KNLA, and KLA) are seen as legitimate, both domestically and internationally. Some movements are seen as illegitimate and unconscionable savagery (RUF in Sierra Leone). It is more likely, however, that a group will be seen as legitimate by its own constituency but not by a majority of outside actors (the Marxist revolutions, most of the radical Islamist

Appendix E

revolutions, and even heinous phenomena like the Hutu genocide of Tutsis in Rwanda). It is also very common for the judgment of legitimacy to change over time with evolving conditions.

E-49. For example, the NPA, FARC, Sendero Luminoso, and the Iranian revolution are examples where the insurgent movement had considerable domestic legitimacy in the early stages only to see their popular legitimacy erode. In all four cases, the standing government was seen as ineffectual, indifferent, corrupt, and repressive. Therefore, in all four cases, the popular support they enjoyed was a manifestation of genuine and widespread popular upheaval. However, that popular appeal was gradually withdrawn. The NPA failed to join in the 1986 People Power Revolution that swept the hated Marcos regime from power, thus severely damaging its credibility. Like the NPA, the FARC lost legitimacy by failing to modify its program after the standing government improved conditions and made deliberate efforts to protect human rights. Sendero attempted a mass popular approach initially and was popular with highland Indian tribes. However, Sendero's resort to fanatical savagery and rigid discipline severely constrained its legitimacy and potential for growth. Khomeini's exploitation of the broad-based opposition to the hated Shah and his usurpation of that coalition to install a radical Islamic regime was a case of using temporary legitimacy to exchange one totalitarian regime with another.

E-50. By contrast, some movements enhance their legitimacy over time. The KNLA has a clearly understandable and long-established domestic grievance against a murderous regime, and enjoys a largely respected international reputation for lawfulness among revolutionary and insurgent movements. While Hezbollah is considered a terrorist organization by many international entities, its domestic legitimacy is built on a solid foundation of intimate integration with the host society, a broad-based and inclusive provision of social services, and an unwavering antipathy to Israel. The Viet Cong represent an example of why legitimacy does not equate to gentleness of methods. The Viet Cong established legitimacy largely through the Maoist method of systematic and relentless elimination of opponents and methodical expansion of its support base. The Viet Cong also benefited from being a component of the side that eventually prevailed in the conflict. Even the FMLN and PIRA increased their legitimacy over time. The FMLN's record of human-rights abuses and failures of governance rivaled that of its opponents. The PIRA squandered its initial popular support by embarking on an undisciplined terrorist bombing campaign that drew repression and hardship onto its constituent supporters. In both cases, however, a turn toward political activity versus continued violence increased achievement of their goals, reduced violence, and correspondingly increased their ultimate domestic and international legitimacy.

EXTERNAL SUPPORT

E-51. The range and type of external support to revolutionary movements and insurgencies varies. The Iranian revolution neither received nor required significant external support, and the support for the Orange Revolution and Solidarity movements was mostly limited to political and moral support. On the other extreme, some movements, such as the Viet Cong, receive massive and continuous systemic external support as a component of overall integrated strategy. It is unsurprising that Marxist movements (Viet Cong, FMLN, FARC, and NPA) usually received assistance from Marxist governments and Islamic radical movements (al-Qaeda, Hezbollah, Hizbul Mujahedeen, and PLO) or those affiliated with Muslim fighters (Chechens, Taliban) are supported by Muslim governments.

E-52. There are a few common external support methods. These include direct funding from supportive states and nonstate organizations where the state or organization understands who the support is going to and for what purpose. By contrast, many organizations fund charities or front organizations for humanitarian or other nonlethal purposes, only to have those funds unwittingly diverted to armed groups. As a matter of religious observance, Muslims are required to donate a portion of their income to the cause of Islamic charity. Some of this money is funneled to revolutionary and insurgent groups instead. Another very common source of funding is voluntary or coerced support from ethnic, religious, or national diasporas. Finally, some support is given in exchange for commercial interests, such as mining concessions or mercenary contracts.

RESPONSE TO GOVERNMENT COUNTERMEASURES

E-53. There are some cases where the revolutionary movement or insurgency prevailed completely over all government countermeasures. In such cases, the opposition to the prevailing government was so widespread that nothing was likely to stop the movement. In Poland, although the government technically had the means to violently suppress Solidarity members, the movement was so widespread that the government realized it could not function without popular support. Although Khomeini adeptly leveraged Islamist grievances to take over the broad-based revolution in Iran, initial opposition to the Shah was insurmountable as it came from all directions of Iranian political life: radical Islamist, moderate Islamist, secularist, monarchist, Marxist, and so on. The Viet Cong was successful in the larger context of communist defeat of the South Vietnamese government and reunification with the North. The KLA was successful in gaining western support for autonomy in the larger context of Yugoslavian breakup and Serbian brutality.

E-54. In some cases, the revolutionary movement was partially successful by transitioning its methods from violent activities to political participation in response to government opposition. The FMLN in El Salvador was unable to prevail over the state government through terrorism and atrocities, guerrilla warfare, or conventional battles. Eventually, this violent stalemate and the war weariness of the population led the FMLN to accept participation in the political process as an alternative to continued violence. Likewise, the Sinn Fein political wing of PIRA eventually achieved more success in political participation than did the ever-more-indiscriminate campaign of bombing.

E-55. In some cases, revolutionary or insurgent movements mistook prevailing strategic conditions and squandered exploitation of government measures. When the People Power Revolution overthrew Marcos in the Philippines and the Aquino government offered amnesty, liberalization, and economic reforms, the NPA's political boycott and continued violent resistance alienated the populace. PIRA's indiscriminate bombing campaign provoked a British policy of unlimited detention and strict population controls which likewise alienated the PIRA's popular base.

E-56. Sometimes, government countermeasures succeeded in defeating the revolutionary movement of insurgency. It has not been uncommon for states (Peru, Colombia, El Salvador, Iran, Sri Lanka, Russia, and so on) to initially overreact in their counterinsurgency programs, typically emphasizing a military-centric approach relying on ultimately self-defeating methods of brutal repression. Sometimes, however, these states eventually prevailed through a combination of socioeconomic initiatives combined with successful intelligence operations and political outmaneuvering. Sendero Luminoso, PIRA, LTTE, EIJ, al-Qaeda, and the Chechens are all examples where government intelligence operations eventually led to the killing or capturing of the movements' strategic leadership. The NPA, FARC, FMLN, and Sendero are examples of movements reduced in effectiveness as the government security forces became more professional and disciplined, and were used within a larger government campaign to address popular grievances.

E-57. Finally, some revolutionary movements or insurgencies have thwarted government responses and—while they have not yet achieved all of their objectives—continue to resist authority. The al-Qaeda and the Taliban have been badly damaged by ongoing counterterrorism and counterinsurgency but continue to fight and enjoy popular support in Muslim countries. MEND has not been defeated for various reasons, including the difficulty of the geographic area, the ineffectiveness of government responses, and the widespread opportunities for corruption and profit amongst both insurgents and greedy government officials. The PLO and Hezbollah have survived decades of government opposition but continue to contest Israel. The KNLA, despite a decades-long genocidal campaign to defeat it, continues to survive as the longest-lived insurgency on the periphery of Burmese control.

This page intentionally left blank.

Appendix F
Unconventional Warfare Campaign Analysis

The following paragraphs discuss a partial list of the areas that should be considered during an analysis of each of the PMESII-PT variables. Some may be potential nodes in each of the systems as well.

POLITICAL SYSTEM POINTS OF ANALYSIS

F-1. The political system points of analysis include the following:
- Leadership (national, regional, local, and core).
- Local workers parties.
- Secret police.
- Informants.
- Security apparatus.
- Detention camps.
- Alliances and external support.
- Regime control of national resource systems.
- Domestic image of omnipotence, omnipresence, and infallibility.

F-2. Political analysis of a foreign country begins with an assessment of the basic principles of government, governmental operations, foreign policy, political parties, pressure groups, electoral procedures, subversive movements, as well as criminal and terrorist organizations. It then analyzes the distribution of political power—whether it is a democracy, an oligarchy, a dictatorship, or has political power devolved to multiple interest groups, such as tribes, clans, or gangs. Analysis must focus on determining how the political system really operates, not the way it is supposed to operate.

BASIC GOVERNMENTAL PRINCIPLES

F-3. The starting point of political analysis is the formal political structure and procedure of a foreign nation. Analysts must evaluate—
- Constitutional and legal systems.
- Legal position of the legislative, judicial, and executive branches.
- Civil and religious rights of the people.
- People's national devotion to constitutional and legal procedures.

GOVERNMENTAL OPERATIONS

F-4. Governments are evaluated to determine their efficiency, integrity, and stability. Information about how the government actually operates or changes its method of operation gives the intelligence user clues about the probable future of a political system. When assessing governmental operations, analysts should consider the following:
- Marked inefficiency and corruption, which differs from past patterns, may indicate an impending change in government.

Appendix F

- Continued inefficiency and corruption may indicate popular apathy or a populace unable to effect change.
- Increased restrictions on the electoral process and on the basic social and political rights of the people may mean the government is growing less sure of its position and survivability.

FOREIGN POLICY

F-5. Analysis of a target country's foreign policy addresses the country's public and private stance toward the United States, foreign policy goals and objectives, regional role, and alliances. Analysts gather foreign policy data from various sources, to include—

- Diplomatic and military personnel.
- Technical collection systems.
- Official foreign government statements.
- Press releases.
- Public opinion polls.
- International businessmen.
- Academic analyses.

POLITICAL PARTIES

F-6. Analysts study special interest parties and groups (for example, labor, religious, ethnic, and industry) to evaluate their—

- Aims.
- Programs.
- Degree of popular support.
- Financial backing.
- Leadership.
- Electoral procedures.

PRESSURE GROUPS

F-7. With few exceptions, most states have some type of formal or informal pressure groups. Examples include political parties, associations, religious or ethnic organizations, labor unions, and even illegal organizations (for example, a banned political party). The analyst must identify these pressure groups and their aims, methods, relative power, sources of support, and leadership. Pressure groups may have international connections and, in some cases, may be almost entirely controlled from outside the country.

ELECTORAL PROCEDURES

F-8. Elections range from staged shows of limited intelligence significance to a means of peaceful, organized, and scheduled revolution. In addition to the parties, personalities, and policies, the intelligence analyst must consider the circumstances surrounding the actual balloting process and changes from the historical norm.

SUBVERSIVE MOVEMENTS

F-9. In many countries there are clandestine organizations or guerrilla groups whose intention is to overthrow or destroy the existing government. When analysts report on subversive movements, they should address—

- Organizational size.
- Character of membership.
- Power base within the society.
- Doctrine or beliefs system.

- Affiliated organizations.
- Key figures.
- Funding.
- Methods of operation.

CRIMINAL AND TERRORIST ORGANIZATIONS

F-10. Criminal organizations in some countries are so powerful that they influence or dominate national governments. Analysts must examine the organization's influence or forceful methods of control. Most terrorist organizations are small, short-lived, and not attached to any government. Analysts should determine if external factors or even the area's government assists the terrorist group.

POLITICAL SYSTEM QUESTIONS

F-11. The following political system questions pertain to the national political structure:
- What is the type of governmental system in place?
    - Where does it draw its legitimacy from?
    - Are the sectors stable or in transition?
    - Does the electoral process affect them?
    - Where do they draw their power?
    - What is the source of their knowledge and intellectual income?
- Who are the leaders?
    - Where do they draw their power from?
    - Does a core bureaucracy staff them?
    - Governmental departments or agencies?
- Who are the key leaders?
- How are they linked within the power network?
- Are the governmental departments or agencies stable or in transition?
    - Are new departments or agencies being created?
    - If so, what is the cause of this transition? (Societal/cultural/educational/technical/economic?)
- What is the source of its workforce?
- Who are the leaders?
- Is it staffed by a core bureaucracy?
    - What skill level?
    - Interagency and departmental dependencies?
    - External dependencies? (Societal/cultural/educational?)

F-12. The following questions pertain to the national political demographics structure:
- Ethnic and religious groups having political power:
    - Are these groups regionalized?
    - How do they exercise political power?
    - What is their legislative representation?
    - Is there a paramilitary structure?
    - How do these ethnic and religious groups wield power within urban society?
    - How do these ethnic and religious groups wield power within rural society?
- Political parties:
    - What are the political parties?
    - Externally or internally supported?
    - Are they associated with ethnic, religious, or cultural groups?

Appendix F

- Who are their leaders? Who are their allies?
- What is their political opposition? Who are their allies?
- Political action groups:
  - Where do they draw their power? (Societal, cultural, technical, economic?)
  - Where do they draw their intellectual capital?
  - What is the source of their leadership?
  - Knowledge?
  - What are their external organs?
  - Expatriate communities?
  - What is their relationship with the government?

F-13. The following questions pertain to the regional political relationships:
- Regional—Nonadversarial and adversarial?
- How are relations maintained—through economics, religion, culture, ideology, common needs?
- International—Nonadversarial and adversarial?
- How are relations maintained—through economics, religion, culture, ideology, common needs?
- Potential allies during a conflict: National resolve to engage in conflict?
- Military resolve to engage in politically motivated action?

F-14. Other considerations include the following:
- Public confidence in government and in society.
- Factionalism or regionalism within the governmental structure.
- Challenges faced by the government.
- Political effects caused by organized groups.
- Government political response to group pressures.
- Political effects upon internal and external security—relates to military.
- Government response to diplomatic overtures.
- National economic goals affecting the political structure.
- Police mechanisms.

MILITARY SYSTEM POINTS OF ANALYSIS

F-15. The military system points of analysis include the following:
- Leadership.
- Intelligence:
  - Signals intelligence.
  - Human intelligence.
- Logistics.
- Command and control.
- Electronic warfare.
- Civil defense.
- Underground facilities.
- Power ventilation access.
- Mobilization.
- Training.
- Stockpiles.
- Army:
  - Artillery.
  - Long-range missile systems.

- Infantry.
- Armor.
- Engineers.
- Mobility.
- Mine-clearing command and control.
- Bridging.
- Countermobility.
- Obstacles.
- Survivability.
* Navy:
  - Surface capabilities.
  - Subsurface (submarine).
  - Remote control vehicles.
  - Mine-laying submarines.
  - SOF platforms.
  - Patrol fleet anti-ship missiles.
  - Coastal defenses.
  - Radar capabilities.
* Air Force:
  - Air-to-ground capabilities.
  - Fixed-wing platforms.
  - Rotary-wing platforms.
  - Air-defense capabilities.
  - Radar and integrated air defense system (IADS).
  - Precision munitions capabilities.
  - Bases (runways, refuel capabilities, ramp space).
* Communications: Missiles (theater/ballistic)/space.
* WMD (research, production, storage, delivery).
* Industrial/technical base (for production and repair of advanced equipment).

ANALYSIS OF THE ADVERSARY'S MILITARY

F-16. The analysis of the adversary's military will focus on its leadership, capabilities, dispositions, and morale/commitment to its government, to include—
* Key military leadership, including their training and previous experience in senior leadership.
* Installations and facilities of a military significance (both primary and secondary purposes).
* Infrastructure in place to support identified installations and force structure.
* Military units, including personnel and chain of command.
* Assigned equipment.
* Current and projected weapons system capabilities.

MILITARY SYSTEM QUESTIONS

F-17. The following military system questions pertain to the military environment:
* Will the national leadership use military means to achieve objectives?
* Does the leadership intend to forge or enhance military ties with another state that poses a threat to regional security or U.S. interests?
* Does the leadership intend to enhance national military capabilities in a way that could be regionally destabilizing?

Appendix F

- Are the national leader's political goals a cause for concern?
- Key leadership—residence, office, wartime command post, telephone, email, political patronage, religious affiliations, ethnic affiliations, personal assets, nonmilitary activities, influences?
- Soldiers—ethnic/religious composition by region of regular forces and elite forces, pay, training, morale, benefits, gripes/issues?
- Capabilities:
    - Equipment imports—what, from whom, where based, points of entry?
    - Support (spare parts, maintenance and operational training)?
    - Indigenous production and assembly?
    - Raw materials, natural resources?
    - Supply—production, movement, storage?
    - Days of supply on-hand of key supplies (ammunition and POL)?
- Transportation—road capacity, primary lines of communication, organic transportation assets?
    - Rail (same as roads)?
    - Water (inland/intracoastal)?
    - Bridges (classification, construction materials, length, bypass)?
    - Tunnels (height/width restrictions, bypass)?
- Organizations:
    - Garrison locations, brigade or larger combat, battalion or larger logistics and sustainment?
    - Naval port facilities, home stations?
- Airfields:
    - Fixed fields?
    - Home station?
    - Associated dispersal/highway strips?
    - Number and type aircraft at base?
- Reconnaissance and surveillance:
    - Assets and capabilities by echelon?
    - National-level/controlled assets?
    - Associated ground stations/downlinks?
    - Centralized processing and dissemination facilities?
    - Center of excellence/HQ for each intelligence discipline?
    - Commercial sources for imagery, dissemination capability, mapping, other?
- Military communications:
    - Fixed facilities?
    - Mobile capabilities?
    - Relay/retransmission sites?
    - Commercial access?
- Integrated air defense:
    - Early warning?
    - Target acquisition and tracking, guidance?
    - Fixed launch sites?
    - Mobile air defense assets?
    - Centralized command and control?
    - Airfields associated with counter-air assets?
    - Airborne warning aircraft (for example, Airborne Warning and Control System)?
    - Electrical power requirements?

- Theater ballistic missile/coastal defense missiles:
  - Fixed launch sites?
  - Mobile assets?
  - Meteorological stations supporting?
  - Command and control decisionmakers?
  - Target acquisition?
  - Target guidance/terminal guidance?
  - Power requirements?
- Weapons of mass effects capabilities:
  - Number and type?
  - Production, assembly, storage, delivery means?
  - Imports required—source and mode of transport?
  - Command and control decisionmaker?
  - Command and control rivalries—personal and interservice?
  - Decisionmaking—dissemination/transmission means (direct or through chain of command)?
- Special capabilities:
  - SOF?
  - WMD?
  - Theater ballistic missile?
  - Human intelligence?
  - Submarines?
- Military situation:
  - Under what conditions does the military execute its missions?
  - Is there internal conflict within the military that could destabilize this country?
  - Are there emerging or increasing rivalries or factionalism within the military?
  - Are there emerging or increasing power struggles within the military?
  - Is there deteriorating morale or increasing dissension within the ranks or in the officer corps?
- Civil-military relations:
  - How loyal is the military to the current regime?
  - Are there cultural or religious factors that might cause frictions and dissension?
  - Are there changes or developments in civil military relations that could destabilize the country?
- Government-military relations:
  - Will the senior military leadership support and defend the government against internal resistance and insurgency?
  - What factors might cause a loss of confidence and/or support?
  - What factors might cause a military coup to occur?
- Civil-military conflict:
  - Is there increasing conflict between the civilian and military leaders?
  - Is there a difference in views between junior and senior leaders toward service to the government? To the peoples/constitution?
  - Is there increasing civil-military conflict over constitutional/legal matters?

- Socio-military conflict:
    - Are there growing tensions/conflicts in socio-military relations that could destabilize the country?
    - Is the military assuming a new internal security role or increasing its involvement in internal security affairs?
- Military activities:
    - Are military operations/activities having an increasingly adverse impact on society?
    - Is the military involved in criminal activity that is contributing to increased tensions/conflict between the military and the public?
- External military threat:
    - Is an external military threat emerging or increasing?
    - Is an adversary engaging in or increasing limited/covert military action?
    - Is an adversary preparing to engage in conventional military action against this country?
    - Is an adversary trying to acquire or is in the process of deploying WMD or advanced weapons?
- Operational status/capability:
    - Are there changes or developments in the military's operational status or capabilities that suggest pending military action?
    - Is there unusual change or a sudden increase in activity levels/patterns?
    - Are there changes or developments in personnel status?
    - Are there significant changes or developments in force capabilities?

ECONOMIC SYSTEM POINTS OF ANALYSIS

F-18. The economic system points of analysis include the following:
- Industry.
- Financial.
- Distribution humanitarian aid.
- Currency.
- Arms exports.
- Corruption/linkages.
- Black market agriculture.
- Drug crops and trafficking.
- Mining.
- Natural resource areas/production.
- Foreign investment.
- Trade linkages.

ANALYSIS OF THE ADVERSARY'S ECONOMY

F-19. Analysis focuses on all aspects of the adversary's economy that have the potential for exploitation. Among these are industrial production, agriculture, services, and armament production. Concentration will be on those elements of the economy that are factors in foreign trade and factors on the internal economy that can have an impact on the political decisionmaking process and popular support for the government. Both the official and underground (black market) economies must be examined.

F-20. Concentration will be on the adversary and the regional and global countries with which it has its major trade and exchange linkages. Certain specific nations and regional economic alliances could be highly dependent upon adversary exports, and the impact upon these must be considered. The focus will be on critical elements of the trading partners that may be exploited and not their economy as a whole. In the economic system, a great deal of information is available from open sources. The initial task is to develop a

baseline of information on the adversary's economy, such as gross domestic product, growth rates, unemployment rates, money supply, economic plans, inflation, and national debt. Analysis may include sources of national wealth:
- Natural resources.
- Products (agriculture and manufacturing).
- Foreign aid.
- Foreign trade.
- Import/export.
- Trading partners.
- Domestic consumption.
- Management of the economy.
- Government role.
- Private sector role.
- Corruption.
- Slush funds, leaders' bank accounts.
- Counterfeiting.

ECONOMIC SYSTEM QUESTIONS

F-21. The following questions pertain to the economic system:
- What are the key indicators of the economic health of the countries of interest (COIs)?
- Which external factors have the most impact upon the economy?
- What areas of the economy are most susceptible to foreign influences and exploitation?
- What is the impact of foreign economic assistance? What would be the impact of its reduction/removal?
- What percentage of the economy should be classified as "black/gray market"?
  - Can the activities in this sector be quantified?
  - Can the activities in this sector be influenced?
- What are the governmental rules on foreign investment? Who do they favor?
- Which nations have the most to gain or lose from damage to, or a collapse of, the economy?
- What are the most likely areas of economic growth?
- Will there be growth in the private sector share of the economy? Who would benefit the most from this change?
- How effective will be steps to diversify the economy?
- What is the inflation rate? To what extent will steps to control inflation be successful?
- Will government subsidies of selected products for domestic use continue? What would be the impact of their reduction/removal?
- What is the anticipated trend in demand for foreign (particularly U.S.) currency?
- What is the prognosis for food production?
  - Are they dependent on imports?
  - Will rationing of essential goods continue?
  - Which items are most likely to be rationed?
- How will demographic factors (for example, birth rate, adult/child ratio, or rural migration to urban areas) affect the economy in the future?
- What is the impact of the drug trade on the overall economy? Regional economies?
- Will imports of military spending/hardware increase?
  - Who are the most likely suppliers?
  - Will these be cash transactions or will a barter system be established?

Appendix F

- What is this nation's standing within the International Monetary Fund and World Bank?
- Is trade with European Union member nations expected to increase? If so, in what specific areas?
- Have any key members of the economic sector leadership been educated in the West or China? If so, have they maintained contacts with their former colleagues?
- Are changes to the current system of state-owned monopolies anticipated? If so, what will be the impact?
- What are the key industries of the state(s)?
- What are the major import/export commodities?
- What is the trade balance? Is this a strength or vulnerability?
- What is the labor situation (for example, unemployment statistics, labor sources, unions, and so on)?
- Who/what are the key government economic leaders/agencies?
- Who are the principal business leaders in the country?

SOCIAL SYSTEM POINTS OF ANALYSIS

F-22. The social system points of analysis include the following:
- Culture/system.
- Personality.
- History.
- Religion command, control, and communications.
- Family ties/tribal linkages.
- Organized crime.
- Impact of local traditions.
- Families (traditional/influential controlling major decisions).

F-23. Analysis must study the way people, particularly the key leadership and natural leaders, organize their day-to-day living, including the study of groups within society, their composition, organization, purposes and habits, and the role of individuals in society. For intelligence purposes, analysts study seven sociological factors. The detailed list should be viewed as a guide for developing the necessary information to develop the sociological systems summary for the target countries.

POPULATION

F-24. Intelligence data derived from censuses and sample surveys describe the size, distribution, and characteristics of the population, including rate of change. Most countries now conduct censuses and publish detailed data. Analysts use censuses and surveys to evaluate an area's population in terms of—
- Location.
- Growth rates.
- Age and sex.
- Structure.
- Labor force.
- Military manpower.
- Migration.

CHARACTERISTICS OF THE PEOPLE

F-25. Analysts study social characteristics to determine their contribution to national cohesion or national disintegration. Social characteristics evaluated by analysts include—
- Social stratification.
- Number and distribution of languages.

- Prejudices.
- Formal and informal organizations.
- Traditions.
- Taboos.
- Nonpolitical or religious groupings and tribal or clan organization idiosyncrasies.
- Social mobility.

PUBLIC OPINION

F-26. Key indicators of a society's goals may be found in the attitudes expressed by significant segments of the population on questions of national interest. Opinions may vary from near unanimity to a nearly uniform scattering of opinion over a wide spectrum. Analysts should sample minority opinions, especially of groups capable of pressuring the government.

EDUCATION

F-27. Analysts concentrate on the general character of education and on the quality of elementary through graduate and professional schools. Data collected for these studies include—
- Education expenditures.
- Relationship between education and other social and political characteristics.
- Education levels among the various components of society.
- Numbers of students studying abroad.
- Extent to which foreign languages are taught.
- Subjects taught in schools.

RELIGION

F-28. Religious beliefs may be a potentially dangerous friction factor for deployed U.S. personnel; this was experienced in the Middle East with Fundamentalist Islamic sects. Understanding those friction factors is essential to mission accomplishment and the protection of friendly forces. Analysts evaluate data collected on an area's religions, which includes—
- Types.
- Size of denominations.
- Growth or decline rates.
- Cooperative or confrontational relationships between religions, the people they represent, and the government.
- Ways the government deals with religious organizations.
- Roles religious groups play in the national decisionmaking process.
- Religious traditions and taboos.

PUBLIC WELFARE

F-29. To evaluate the general health of a population, analysts must identify—
- Health delivery systems.
- Governmental and informal welfare systems.
- Social services provided.
- Living conditions.
- Social insurance.
- Social problems that affect national strength and stability (for example, divorce rate, slums, drug use, or crime) and methods of coping with these problems.

Appendix F

NARCOTICS AND TERRORISM TOLERANCE

F-30. A population's level of tolerance for narcotics and terrorist activities depends on the relations between these organizations and the population as a whole. Analysts should determine if the tolerance is a result of the huge sums of money traffickers pump into the economy or a result of traffickers' use of force. Terrorists may be accepted and even supported by the local populace if they are perceived to be working for the good of the local people. The intelligence analyst must evaluate the way these organizations operate.

SOURCES

F-31. Because of the nature of the social focus area, the preponderance of information is envisioned to be open source. The initial task is to develop a baseline of information on the target nation. Basic data will be collected and analyzed. Numerous studies, sponsored by the USG as well as academic treatises are available. A more difficult problem will be making the essential linkages within the sociological area and with other focus areas, particularly political and economic.

SOCIAL SYSTEM QUESTIONS

F-32. The following questions pertain to the social system:
- What are the general perceptions of social stability?
- Who are the population's most respected figures, why are they so respected, and how do they maintain the public focus?
- What are the government's most effective tools for influencing the masses?
- What dominant areas of society are emerging and causing instability or areas of conflict?
    - Are any of these areas linked to political factors?
    - Are there ethnic/racial factors?
- What are the predominant economic areas that are contributing to, promoting, or exacerbating social instability?
- How can interrelationships be established between religious and ethnic minorities in the COI? How can these relationships be effectively manipulated to affect a desired outcome?
- What are perceptions of public safety primarily attached to?
    - How is the level of violence defined by society?
    - What elements may make it appear excessive?
- What psychological effects does an increased level of violence have on a person's notion of safety?
- What are the effects of increased criminal activity on the family, the town, the region, and nationally?
- How can the coalition increase the psychological perception that the global economy is surpassing the COI?
- How can the coalition stimulate the notion that the government is failing to provide for basic elements, or is slow to produce results?
- What are the adverse effects of increased organized criminal activity upon society by industrial component?
    - Is there white collar or financial crime?
    - Are there issues with drugs and drug smuggling?
- Proliferation of weapons: What are the types of weapons and to whom are they going?
- Gang-related activity: Is there a predominant ethnic group asserting themselves in this arena, and are they utilizing any particularly violent tactics to assert themselves?
- What are the significant effects of increased public health problems?
    - What public health issues have increased?
    - How effective is the government?

- How extensive is the division of wealth between ethnic and religious groups and what is their potential for promoting tension or conflict?
- What are the effects of environmental problems having on society?
- What are the key groups adversely affected by increasing poverty rates?
- What primary tools are used by the government for influencing the masses?
  - How do the masses validate information obtained by the government?
  - Do they feel they need to validate information?
- Who are the key opposition leaders?
  - How do they influence the masses?
  - How are they funded and by whom are they primarily funded?
- Who are the key opposition groups?
  - How do they influence the masses?
  - How are they funded and by whom are they primarily funded?
  - Are there any common themes to unite them or areas that may divide them?
- How do opposition groups recruit?
  - Do they target a specific social group?
  - Is there a hierarchical structure?
  - How are members dismissed from the ranks?
- How do these groups affect one another?
  - How do they affect similar groups in neighboring countries?
  - Do they have external support?
- What are each faction's mechanisms for influencing the others?
  - How do they communicate officially and unofficially?
  - What factions are armed?
  - Where do they get their weapons?
- Are acts of civil disobedience increasing?
  - Is the level of violence employed by the government to quell civil disobedience increasing?
  - Are acts of vigilantism on the rise?
  - How are disturbances quelled?
  - What tools are brought to bear?
- What consumer goods are most valued by the COI's populace?
  - Who controls supply?
  - How are they networked?
  - Is there any increase in a particular product?
- What are the "hot button" issues dividing the various factions of the society?
  - What networks and mediums can be used to subvert and confuse each faction?
  - What are the capabilities of regional allies to polarize these factions?
- How are rumors spread most effectively?
- What is the social perception of the military's ability to meet that threat?
  - What is the state's ability to meet the threat?
  - What is the state's ability to provide overall security in a micro/macro context?
- How are troops conscripted?
  - What are the incentives for service?
  - What unofficial groups/associations exist within the military?
  - How do they recruit or dismiss people?
- Is criminal behavior increasing within the military?
- What types of criminal activity occur within the military?

Appendix F

- What is the hierarchal structure of the military?
  - Is there a dominant ethnic group assuming more leadership roles?
  - What ethnic groups stay the most connected in the military?
  - Which groups are more apt to include outsiders?
- Which ethnic and religious minorities feel the most repressed?
  - How do they express their discontent?
  - Do any organizations exist to channel their feelings?
  - How responsive do they feel the government is to their issues?
- How does the population view outside assistance?
  - How likely is the government to ask for assistance?
  - How is the need for assistance determined?
- How are relief organizations viewed within the country?
  - Are they busy?
  - How effective are they at solving problems and meeting the needs of those they serve?
- Are there problems with immigrant flows?
- How are refugees treated?
- What consumer goods are in short supply?
  - How are those goods brought to market?
  - Who controls the flow of such goods?
  - Is there a dominant ethnic group controlling the flow?
  - How effective is the black market in producing hard-to-obtain goods?
- What goods dominate the black market?
  - Who are the primary producers and end receivers of goods?
  - Is there a particular group emerging as the leader of the black market?
- How are minority laborers networked with minority leaders?
  - What are the links between labor groups and minority activists?
  - What ethnic groups compose the majority of the skilled labor force?
  - How is skilled labor kept from going abroad?

INFRASTRUCTURE SYSTEM POINTS OF ANALYSIS

F-33. The infrastructure system points of analysis include the following:
- Transportation:
  - Rail.
  - Trains.
  - Bridges.
  - Tunnels.
  - Switches.
  - Road.
  - Ship/boat.
  - Dam locks.
  - Air.
- Communications:
  - Military networks.
  - Radio telephone.
  - Teletype fiber satellite.
  - Visual.

- Civilian.
- Radio telephone.
- Television speakers.
- Signs.
- Energy/power:
  - Coal.
  - POL.
  - Hydro.
  - Nuclear.
  - Water.

F-34. Infrastructure analysis focuses on the quality and depth of the physical structures that support the people and industry of the state. In developed countries, it is the underlying foundation or basic systems of a nation state, generally physical in nature and supporting/used by other entities (for example, roads, telephone systems, and public schools).

INFRASTRUCTURE SYSTEM QUESTIONS

F-35. The following questions pertain to the infrastructure system:
- Lines of communications:
  - Where are the key ports, airfields, rail terminals, roads, railroads, inland waterways located?
  - Where are key bridges, tunnels, switching yards, scheduling/control facilities, depots/loading stations, switching yards, and so on?
- Electrical power:
  - Where are power plants, transformer stations, and relay and power transmission lines located?
  - Where are the key substations, switching stations, and line junctures?
- Potable water:
  - Where are the water treatment plants, wells, desalination, bottling plants, and pumping stations?
  - Where are the key pumping stations, control valves, and distribution line junctures?
- Telecommunications:
  - What are the location and architecture of the domestic telephone system, cable, fiber-optic, microwave, Internet, and cell phone networks and satellite stations?
  - Where are the key control points and junctures?
- Petroleum and gas:
  - Where are the gas and petroleum fields, gathering sites, pumping stations, storage areas, refineries and distribution lines?
  - Where are the key pumping stations, control valves, and distribution junctures?
- Broadcast media:
  - What are the location, frequency, power, and radius of effective range (coverage) of the amplitude modulation/frequency modulation radio and television stations?
  - Where are the studios, antenna, and relay towers located?
  - How are they powered?
  - Where are the key control points and junctures?
- Public health:
  - What are the locations of the hospitals and clinics?
  - Are they adequately staffed, supplied, and equipped?
  - Is the equipment well maintained?

Appendix F

- Is the staff well trained?
- Do they depend on foreign or domestic sources for their supplies, medications, and spare equipment parts?
- Where are the key control points and junctures?
- Schools:
  - What are the locations of the public, private, and religious primary and secondary schools and universities?
  - Where are the key control points and junctures?
- Public transportation:
  - What are the public (bus/streetcar/taxi) transportation routes?
  - Where are the key control points and junctures?
- Sewage collection and treatment:
  - Where are the collection systems, pumping stations, treatment facilities, and discharge areas located?
  - Where are the key control points and junctures?

OTHER COMMON INFRASTRUCTURE QUESTIONS

F-36. Other common infrastructure questions include the following:
- How are key facilities linked? (Physically, electronically, and so on.)
- What are the key nodes?
  - Where are they?
  - Where are the disabling yet nonlethal/nondestructive infrastructure nodes?
  - What are their alternates?
  - How are they linked to the key facilities and each other?
- Are there indigenous capabilities?
  - What indigenous capabilities could be used?
  - How are they linked and organized?
- What are the critical nodes?
  - What is the security surrounding the nodes?
  - What is the security posture at these facilities?
- Who controls the forces?
  - How are security forces/police/paramilitary networked?
  - What training do they receive?
  - What is their level of proficiency?
  - Are they augmented as alert status (national or local) changes?
- What are the ground/naval/air defense capabilities at or near these facilities?
  - How are they networked?
  - What groups are likely to conduct industrial sabotage?
  - How are they tasked, linked, supported?
- Who owns and who controls the infrastructure?
- Who owns and/or controls the aforementioned entities?
- Is ownership by private, corporate, or governmental entities?
- What organizations have regulatory oversight/control?
- What is the capability to repair damage to the system and restore it to service?
- Is maintenance and repair an integral part of the organization?
- What are their capabilities and limitations?
- Which contractors are normally used and for what purpose?

- Are repair/restore materials readily available or is there a long lead time for critical supplies and components?
- Who are the key engineering contractors for these facilities?
- Can/will they share plans, blueprints, schematics, and so on?
- What would be the second-order effects of influencing the infrastructure?

INFORMATION SYSTEM POINTS OF ANALYSIS

F-37. The information system points of analysis include the following:
- Education.
- Propaganda: Inside country; outside country.
- Newspapers/magazines information technologies.
- Radio.
- Television.
- Internet.
- Informal transmissions (word of mouth/rumor).
- Cyberspace information systems.

ANALYSIS OF INFORMATION SYSTEMS

F-38. Analysis of information systems and operations includes the following:
- Telecommunications capabilities and level of sophistication, teledensity rates, radio and television broadcast coverage, including television, landline, cellular, Internet, radio, and so on.
- Interconnectivity of communications via integrated services digital network, fiber optic, satellite, and microwave.
- Primary nodes and trunks of telecommunications infrastructure, including government, nongovernment, citizen, and military use of IO.
- Knowledge of COI key leaders' style and decisionmaking habits, advisors' perception, and cultural influences.
- Understanding governmental use of media influence, public affairs, and CA interrelationships.
- Knowledge of military, nongovernmental organization, and law enforcement interrelationships.
- Understanding of effects on adversary under psychological, computer network attack and defense, electronic warfare, and space operations.
- Locations and purpose of physical infrastructure of communications and broadcast towers, cables, and supporting operations centers included within the infrastructure focus.
- Development of and use of computer network operating systems, information technology industry skill sets, and software applications.
- Media affiliations, perceptions, and sympathies, to include censorship and self-censorship in news and entertainment print and broadcast industries.

INFORMATION SYSTEM QUESTIONS

F-39. The following questions pertain to the information system:
- How effective are the COI's network defense capabilities?
- What reactions could be expected following an incident?
- What recovery procedures are routinely exercised?
- What is the organizational structure of the telecommunications industry?
- How effective is the COI at managing physical security of infrastructure and implementing network security practices?
- What interrelationships exist between civil law enforcement, military, commercial, and nongovernmental agencies that would enhance the COI's response to an emergency?

Appendix F

- What redundancies exist within the COI's network to eliminate or reduce network down time?
  - Cellular, satellite, landline, power back up?
  - How effective is their exchange, backbone, architecture in providing redundancies?
- What would cause a slowdown of the COI's network?
- In what ways can the effect be localized (geographic, logic, by agency, and so on)?
- What bandwidth issues exist within the COI's communications industry?
- How well, and in what ways, does the government manage its allocation?
- What type of operations security practices does the COI routinely exhibit to deny exploitation?
- In what ways have military/civil/corporate operations centers improved their practices/tactics in keeping with the COI's technological improvements?
- Do they rely more heavily on computers/cellular/networks than in the past?
- What are the indicators, if they exist, that the COI has developed a more focused vision and strategic plan for using technology than it had in the late 1990s?
- What effect has technology had on productivity, transportation, logistics, and so on in government, commerce, corporate, and private sectors?
- How does the COI perceive their use of technology from a governmental perspective?
  - From the citizens' perspective?
  - Military?
  - Business?
  - Legal?
  - Law enforcement?
  - NGOs?
- What is known about the COI's assessment of Blue network vulnerabilities and defense measures?
- Do regional and neighboring countries or satellite broadcasts (television, radio, and Internet) have an audience in the COI's population?
- Which broadcasts are popular with citizens and what are the audience's demographic and statistic data?
- What programs or broadcasts are popular with minority political parties, resistance movements, academia, and so on?
- What is the topology design the COI networks utilize?
- Which exchanges and trunks are collocated within government-controlled facilities?
- Are government-commercial partnerships used to provide network services?
- What is known of current and planned technology projects:
  - Fiber-optic cabling?
  - Integrated services digital network access expansion?
  - Satellite leases and launches?
- What is the operational status and capability of COI's low-earth orbit satellites?
- What Internet domains are accessible to the population?
- Is reliable language interpretation software available?
- What licenses does the government require for Web hosting?
- What governmental directives address network security in supporting national security objectives?
- What messages might be effective in the COI?
- What themes are prevalent in the media?
- What advances in communications technology have enabled improvements in military hardware employment?
- How is telecommunications technology used in law enforcement operations?

- To what degree and direction are telecommunications infrastructure investments impacting military readiness?
- What is the state of international telecommunications connectivity to the COI?
- Which current telecommunications and Internet security operations have been exercised?
- Is there a national crisis action plan?
- What practices and policies does the government use in monitoring information-related media (television, radio, Internet, and so on)?
- What enforcement methods have been employed?
- Which print media and on-line content do citizens turn to for news or entertainment?
- Do censorship policies or self-censorship trends exist in the COI?
- Is there a market and distribution pipeline for recorded or intercepted news or entertainment programs?
- In what ways does law enforcement interact in this market?
- What is known about the COI's network operating systems?
- What information technology skill sets are known to be in high demand?
- Is there a prevalence of—
  - Software piracy?
  - Counterfeiting?
  - Drug smuggling?
  - Organized crime?
  - Identity theft?

PHYSICAL ENVIRONMENT POINTS OF ANALYSIS

F-40. The physical environment points of analysis include the following:
- Terrain, to include—
  - Observation and fields of fire.
  - Avenues of approach.
  - Key terrain.
  - Obstacles.
  - Cover and concealment.
  - Land forms.
  - Vegetation.
  - Terrain complexity.
  - Mobility classification.
- Natural hazards.
- Climate.
- Weather, to include—
  - Precipitation.
  - High-temperature heat index.
  - Low-temperature wind chill.
  - Wind.
  - Visibility.
  - Cloud cover.
  - Relative humidity.

Appendix F

F-41. Physical environment questions include the following:
- What are the natural features in the area?
- What operational impact do rivers have?
- What operational impact do mountains have?
- What are the manmade features in the area?
- What is the operational impact of cities?
- What is the operational impact of airfields?
- What is the operational impact of bridges?

TIME POINTS OF ANALYSIS

F-42. The time points of analysis include the following:
- Cultural perception of time.
- Information offset.
- Tactical exploitation of time.
- Key dates, time periods, or events.

F-43. Time questions include the following:
- What are the time schedules of operationally significant adversary activities?
- How does the environment in the area impact friendly time schedules to assemble, deploy, and maneuver units in relationship to the adversary?

Glossary

SECTION I – ACRONYMS AND ABBREVIATIONS

ADRP	Army doctrine reference publication
AFSOC	Air Force Special Operations Command
ALE	Army special operations forces liaison element
AO	area of operations
AOR	area of responsibility
ARSOF	Army special operations forces
ASCC	Army Service component command
ATP	Army techniques publication
CA	Civil Affairs
CAO	Civil Affairs operations
CARVER	criticality, accessibility, recuperability, vulnerability, effect, and recognizability
CCIR	commander's critical information requirement
CJCS	Chairman of the Joint Chiefs of Staff
CJSOTF	combined joint special operations task force
CMO	civil-military operations
CMOC	civil-military operations center
COA	course of action
COI	country of interest
CONOPS	concept of operations
DA	Department of the Army
D-day	unnamed day on which operations commence or are scheduled to commence
DLPT	Defense Language Proficiency Test
DOD	Department of Defense
DODI	Department of Defense instruction
DOS	Department of State
EIJ	Egyptian Islamic Jihad
FARC	Fuerzas Armada Revolucionarias de Colombia
FLN	Front de Liberation Nationale
FM	field manual
FMLN	Farabundo Marti Frente Para la Liberacion Nacional
GCC	geographic combatant commander
GTA	graphic training aid
HQ	headquarters
IADS	integrated air defense system
IAW	in accordance with

Glossary

IGO	intergovernmental organization
IIA	inform and influence activities
IO	information operations
IPR-C	in-progress review—concept
JFC	joint force commander
JFLCC	joint force land component commander
JFSOCC	joint force special operations component commander
JIIM	joint, interagency, intergovernmental, and multinational
JOPES	Joint Operation Planning and Execution System
JOPP	joint operation planning process
JP	joint publication
JPG	joint planning group
JRSOI	joint reception, staging, onward movement, and integration
JSOA	joint special operations area
JSOTF	joint special operations task force
JTF	joint task force
KDP	Kurdistan Democratic Party
KLA	Kosovo Liberation Army
KNLA	Karen National Liberation Army
LTTE	Liberation Tigers of Tamil Eelam
MACV-SOG	Military Assistance Command, Vietnam—Studies and Observation Group
MARSOC	Marine Corps Forces Special Operations Command
MEND	Movement for the Emancipation of the Niger Delta
METT-TC	mission, enemy, terrain and weather, troops and support available-time available and civil considerations
MIS	Military Information Support
MISO	Military Information Support Operations
MOS	military occupational specialty
NAVSPECWARCOM	Naval Special Warfare Command
NDS	National Defense Strategy
NGO	nongovernmental organization
NMS	National Military Strategy
NPA	New People's Army
NSS	National Security Strategy
OGA	other government agency
OPCON	operational control
OPLAN	operation plan
OPORD	operation order
PE	preparation of the environment
PIRA	Provisional Irish Republican Army

Glossary

PLO	Palestine Liberation Organization
PMESII	political, military, economic, social, information, and infrastructure
PMESII-PT	political, military, economic, social, information, infrastructure, physical environment, and time
POL	petroleum, oils, and lubricants
PSYACT	psychological action
PSYOP	Psychological Operations
PUK	Patriotic Union of Kurdistan
RUF	Revolutionary United Front
SB(SO)(A)	Sustainment Brigade (Special Operations) (Airborne)
SecDef	Secretary of Defense
SF	Special Forces
SFODA	Special Forces operational detachment A
SO	special operations
SOE	Special Operations Executive (UK)
SOF	special operations forces
SOTF	special operations task force
TC	training circular
TSOC	theater special operations command
U.S.	United States
USAFRICOM	United States Africa Command
USAID	United States Agency for International Development
USAJFKSWCS	United States Army John F. Kennedy Special Warfare Center and School
USASOC	United States Army Special Operations Command
USC	United States Code
USCENTCOM	United States Central Command
USEUCOM	United States European Command
USFIP	United States Forces in the Philippines
USG	United States Government
USPACOM	United States Pacific Command
USSOCOM	United States Special Operations Command
USSOUTHCOM	United States Southern Command
UW	unconventional warfare
WMD	weapons of mass destruction
WWII	World War II

SECTION II – TERMS

* area command
    In unconventional warfare, the irregular organizational structure established within an unconventional warfare operational area to command and control irregular forces advised by Army Special Forces.

auxiliary
    For the purpose of unconventional warfare, the support element of the irregular organization whose organization and operations are clandestine in nature and whose members do not openly indicate their sympathy or involvement with the irregular movement. (ADRP 3-05)

cache
    A source of subsistence and supplies, typically containing items such as food, water, medical items, and/or communications equipment, packaged to prevent damage from exposure and hidden in isolated locations by such methods as burial, concealment, and/or submersion, to support isolated personnel. (JP 3-50)

clandestine operation
    An operation sponsored or conducted by governmental departments or agencies in such a way as to assure secrecy or concealment. A clandestine operation differs from a covert operation in that emphasis is placed on concealment of the operation rather than on concealment of identity of the sponsor. In special operations, an activity may be both covert and clandestine and may focus equally on operational considerations and intelligence-related activities. (JP 3-05.1)

covert operation
    An operation that is so planned and executed as to conceal the identity of or permit plausible denial by the sponsor. (JP 3-05)

denied area
    An area under enemy or unfriendly control in which friendly forces cannot expect to operate successfully within existing operational constraints and force capabilities. (JP 3-05)

* government-in-exile
    A government that has been displaced from its country, but remains recognized as the legitimate sovereign authority.

* guerrilla
    An irregular, predominantly indigenous member of a guerrilla force organized similar to military concepts and structure in order to conduct military and paramilitary operations in enemy-held, hostile, or denied territory. Although a guerrilla and guerrilla forces can exist independent of an insurgency, guerrillas normally operate in covert and overt resistance operations of an insurgency.

* guerrilla base
    A temporary site where guerrilla installations, headquarters, and some guerrilla units are located. A guerrilla base is considered to be transitory and must be capable of rapid displacement by personnel within the base.

guerrilla force
    A group of irregular, predominantly indigenous personnel organized along military lines to conduct military and paramilitary operations in enemy-held, hostile, or denied territory. (JP 3-05)

guerrilla warfare
    Military and paramilitary operations conducted in enemy-held or hostile territory by irregular, predominantly indigenous forces. See also **unconventional warfare**. (JP 3-05.1)

Glossary

insurgency
The organized use of subversion and violence by a group or movement that seeks to overthrow or force change of a governing authority. Insurgency can also refer to the group itself. (JP 3-24)

intelligence operations
The variety of intelligence and counterintelligence tasks that are carried out by various intelligence organizations and activities within the intelligence process. (JP 2-01)

intelligence preparation of the battlespace
The analytical methodologies employed by the Services or joint force component commands to reduce uncertainties concerning the enemy, environment, time, and terrain. Intelligence preparation of the battlespace supports the individual operations of the joint force component commands. See also **joint intelligence preparation of the operational environment**. (JP 2-01.3)

joint intelligence preparation of the operational environment
The analytical process used by joint intelligence organizations to produce intelligence estimates and other intelligence products in support of the joint force commander's decisionmaking process. It is a continuous process that includes defining the operational environment; describing the impact of the operational environment; evaluating the adversary; and determining adversary courses of action. (JP 2-01.3)

low-visibility operations
Sensitive operations wherein the political-military restrictions inherent in covert and clandestine operations are either not necessary or not feasible; actions are taken as required to limit exposure of those involved and/or their activities. Execution of these operations is undertaken with the knowledge that the action and/or sponsorship of the operation may preclude plausible denial by the initiating power. (JP 3-05.1)

* mission support site
A preselected area used as a temporary base or stopover point. The mission support site is used to increase the operational range within the joint special operations area.

nonconventional assisted recovery
Personnel recovery conducted by indigenous/surrogate personnel that are trained, supported, and led by special operations forces, unconventional warfare ground and maritime forces, or other government agencies' personnel that have been specifically trained and directed to establish and operate indigenous or surrogate infrastructures. Also called **NAR**. (JP 3-50)

operational environment
A composite of the conditions, circumstances, and influences that affect the employment of capabilities and bear on the decisions of the commander. (JP 3-0)

* pilot team
A deliberately structured composite organization comprised of Special Forces operational detachment members, with likely augmentation by interagency or other skilled personnel, designed to infiltrate a designated area to conduct sensitive preparation of the environment activities and assess the potential to conduct unconventional warfare in support of U.S. objectives.

preparation of the environment
An umbrella term for operations and activities conducted by selectively trained special operations forces to develop an environment for potential future special operations. (JP 3-05)

raid
An operation to temporarily seize an area in order to secure information, confuse an adversary, capture personnel or equipment, or to destroy a capability culminating with a planned withdrawal. (JP 3-0)

Glossary

resistance movement
    An organized effort by some portion of the civil population of a country to resist the legally established government or an occupying power and to disrupt civil order and stability. (JP 3-05)

* shadow government
    Governmental elements and activities performed by the irregular organization that will eventually take the place of the existing government. Members of the shadow government can be in any element of the irregular organization (underground, auxiliary, or guerrilla force).

subversion
    Actions designed to undermine the military, economic, psychological, or political strength or morale of a governing authority. See also **unconventional warfare**. (JP 3-24)

subversive political action
    A planned series of activities designed to accomplish political objectives by influencing, dominating, or displacing individuals or groups who are so placed as to affect the decisions and actions of another government. (JP 1-02)

unconventional assisted recovery
    Nonconventional assisted recovery conducted by special operations forces. (JP 3-50)

unconventional assisted recovery coordination cell
    A compartmented special operations forces facility, established by the joint force special operations component commander, staffed on a continuous basis by supervisory personnel and tactical planners to coordinate, synchronize, and de-conflict nonconventional assisted recovery operations within the operational area assigned to the joint force commander. (JP 3-50)

unconventional assisted recovery mechanism
    A recovery mechanism developed and managed by special operations forces. (DODI 2310.6)

unconventional assisted recovery team
    A designated special operations forces unconventional warfare ground or maritime force capable of conducting unconventional assisted recovery with indigenous or surrogate forces. (DODI 2310.6)

unconventional warfare
    Activities conducted to enable a resistance movement or insurgency to coerce, disrupt, or overthrow a government or occupying power by operating through or with an underground, auxiliary, and guerrilla force in a denied area. Also called **UW**. (JP 3-05)

underground
    A cellular covert element within unconventional warfare that is compartmentalized and conducts covert or clandestine activities in areas normally denied to the auxiliary and the guerrilla force. (ADRP 3-05)

References

REQUIRED PUBLICATIONS
These documents must be available to intended users of this publication.

None.

RELATED PUBLICATIONS
These documents contain relevant supplemental information.

ARMY FORMS
DA forms are available on the Army Publishing Directorate Web site (http://armypubs.army.mil).

DA Form 2028 (Recommended Changes to Publications and Blank Forms).

ARMY PUBLICATIONS
DA publications are available on the Army Publishing Directorate Web site (http://armypubs.army.mil).

ADRP 1-02, *Operational Terms and Military Symbols*, 31 August 2012.
ADRP 3-05, *Special Operations*, 31 August 2012.
ADRP 4-0, *Sustainment*, 31 July 2012.
ADRP 5-0, *The Operations Process*, 17 May 2012.
ATP 3-05.40, *Special Operations Sustainment*, 3 May 2013.
FM 3-05.70, *Survival*, 17 May 2002.
FM 3-05.220, *(S/NF) Special Forces Advanced Special Operations (U)*, 31 January 2007.
FM 3-05.231, *Special Forces Personnel Recovery*, 13 June 2003.
FM 3-05.701, *(C) Army Special Operations Forces Resistance and Escape (U)*, 1 March 2007.
FM 3-18, *Special Forces Operations*, 5 March 2012.
FM 3-24, *Counterinsurgency*, 15 December 2006.
FM 3-53, *Military Information Support Operations*, 4 January 2013.
FM 27-10, *The Law of Land Warfare*, 18 July 1956.
GTA 41-01-005, *Religious Factors Analysis*, 1 January 2008.
TC 18-01, *Special Forces Unconventional Warfare*, 28 January 2011.
TC 31-16, *(S/NF) Special Forces Guide to Preparation of the Environment (U)*, 6 June 2007.

DEPARTMENT OF DEFENSE PUBLICATIONS
DOD publications are available on the Defense Technical Information Center Web site (http://www.dtic.mil/whs/directives/corres/dir.html).

DOD Instruction 2310.6, *Non-Conventional Assisted Recovery in the Department of Defense*, 13 October 2000.
DOD National Defense Strategy, Washington DC, June 2008.

References

JOINT PUBLICATIONS

Joint publications are available on the Defense Technical Information Center Web site (http://www.dtic.mil/doctrine/new_pubs/jointpub.htm).

JP 1-02, *Department of Defense Dictionary of Military and Associated Terms*, 8 November 2010.

JP 2-01, *Joint and National Intelligence Support to Military Operations*, 5 January 2012.

JP 2-01.3, *Joint Intelligence Preparation of the Operational Environment*, 16 June 2009.

JP 3-0, *Joint Operations*, 11 August 2011.

JP 3-05, *Special Operations*, 18 April 2011.

JP 3-05.1, *Joint Special Operations Task Force Operations*, 26 April 2007.

JP 3-10, *Joint Security Operations in Theater*, 3 February 2010.

JP 3-24, *Counterinsurgency Operations*, 5 October 2009.

JP 3-50, *Personnel Recovery*, 20 December 2011.

JP 4-0, *Joint Logistics*, 18 July 2008.

JP 5-0, *Joint Operation Planning*, 11 August 2011.

OTHER PUBLICATIONS

Crossett Edward, ed., *Casebook on Insurgency and Revolutionary Warfare, Volume 2: 1962–2009*, Johns Hopkins University, 2 June 2011.

Donovan, William J., *Simple Sabotage Field Manual*, Office of Strategic Services, Washington, DC, 17 January 1944.

Gorka, Sebastian, *Jihadist Ideology: The Core Texts*, transcript of paper delivered at Westminster Institute, 5 October 2010.

Heilbrunn, Otto, *Partisan Warfare*, Praeger, New York, 1962.

Hogan, Jr., David W., *U.S. Army Special Operations in World War II*, Department of the Army Center for Military History Publication 70-42, Washington, DC, 1992.

Lawrence, T.E., *Seven Pillars of Wisdom: A Triumph*, Oxford, 1922.

Lawrence, T.E., *Twenty-Seven Articles*, The Arab Bulletin, Cambridge, 20 August 1917.

Li Tso-peng, *Strategy: One Against Ten—Tactics: Ten Against One: An Exposition of Comrade Mao Tse-tung's Thinking on the Strategy and Tactics of the People's War*, Hong Qi (Red Flag) Volume No. 23, Foreign Languages Press, Peking, 1966.

Mao Tse-tung, *Selected Military Writings*, Foreign Languages Press, Peking, 1968.

Molnar, Andrew, *Undergrounds in Insurgent, Revolutionary and Resistance Warfare*, Special Operations Research Office, American University, Washington DC, 1963.

The National Military Strategy of the United States of America, 2011: Redefining America's Military Leadership, 8 February 2011.
<http://www.jcs.mil/content/files/2011-02/020811084800_2011_NMS_-_08_FEB_2011.pdf>

The National Security Strategy, Washington, DC, May 2010.
<http://www.whitehouse.gov/sites/default/files/rss_viewer/national_security_strategy.pdf>

Stanton, Shelby, *Special Forces at War: An Illustrated History, Southeast Asia 1957–1975*, Zenith Press, 2008.

Tierney, John J, *Chasing Ghosts: Unconventional Warfare in American History*, Potomac Books, 2006.

Title 10, USC, *Armed Forces*.
<http://uscodebeta.house.gov/browse/prelim@title10&edition=prelim>

Section 167, *Unified Combatant Command for Special Operations Forces.*
<http://uscodebeta.house.gov/browse/prelim@title10/subtitleA/part1/chapter6&edition=prelim>

Title 50, USC, *War and National Defense.*
<http://uscodebeta.house.gov/browse/prelim@title50&edition=prelim>
Section 413, *General Congressional Oversight Provisions,* 3 January 2012.
<http://uscodebeta.house.gov/browse/prelim@title50/chapter15/subchapter3&edition=prelim>

U.S. Army War College Campaign Planning Handbook, Carlisle, PA, May 2011.
<http://www.carlisle.army.mil/usawc/dmspo/Publications/Publications.htm>

USSOCOM Directive 525-5, *(S/NF) Advanced Special Operations (U),* 19 August 2004.
(Copies may be requested from: USAJFKSWCS, SF Directorate,
ATTN: AOJK-CDI-SFD, 3004 Ardennes Street, Stop A, Fort Bragg, NC 28310-9610.)

USSOCOM Directive 525-16, *(S/NF) Preparation of the Environment (U),* 21 August 2007.
(Copies may be requested from: USAJFKSWCS, SF Directorate,
ATTN: AOJK-CDI-SFD, 3004 Ardennes Street, Stop A, Fort Bragg, NC 28310-9610.)

USSOCOM Directive 525-21, *(S) Personnel Recovery (U),* 29 May 2009.
(Copies may be requested from: USAJFKSWCS, SF Directorate,
ATTN: AOJK-CDI-SFD, 3004 Ardennes Street, Stop A, Fort Bragg, NC 28310-9610.)

USSOCOM Directive 525-89, *(S/NF) Unconventional Warfare (U),* 31 May 2012.
(Copies may be requested from: USAJFKSWCS, SF Directorate,
ATTN: AOJK-CDI-SFD, 3004 Ardennes Street, Stop A, Fort Bragg, NC 28310-9610.)

This page intentionally left blank.

Index

A

agitator, 3-3, 3-6, 3-12
Air Force Special Operations Command, 1-3, A-4
airborne operations, 3-35, 3-36
ambushes, 2-4, 2-23, 3-23, 3-25 through 3-29, 3-32, D-2, D-3, E-8
amphibious operations, 3-36
area commands, 2-22
area complex, 2-11, 2-22 through 2-26
auxiliary, 1-1, 1-4, 2-10 through 2-12, 2-16 through 2-21, 2-23 through 2-26, 3-20, 3-23, 3-24, 3-28, 3-29, 3-32, 3-35, 3-37, 4-7, 5-3, A-2 through A-4, C-1, C-5 through C-8

B

buildup, 2-8, 2-10, 2-11, 2-20, 2-26, 4-3, 4-7, B-4, C-3

C

campaign planning, v, 1-1, 3-2, 4-2, 4-4, 5-1, 5-2, 5-4 through 5-6, 5-16, 5-23, A-1, A-3 through A-5
Civil Affairs, 1-9, 2-8 through 2-13, 2-15, 3-2, 3-8, 3-38, 4-1 through 4-3, 5-27, B-3, F-17
Civil Affairs operations, 2-13, 4-2, 4-3, 5-27, A-2
civil disobedience, 3-7 through 3-9, C-1, F-13
civil-military operations, 2-10, 4-2, 4-3, 5-25, 5-27, A-2
civil-military operations center, 2-13
counterguerrilla operations, 2-5, 3-32, 3-37, 3-38

D

demonstrations, 2-6, 3-3, 3-5, 3-7, 3-9 through 3-11, 3-13, 3-20, E-7
disturbance leader, 3-12

F

feasibility assessment, 1-8, 2-9, 2-15

G

government-in-exile, vi, 2-21, 3-37, 4-3
guerrilla base, vi, 2-23, 2-25, 2-26, 3-32
guerrilla hospital, 2-26
guerrilla warfare, 2-7, 3-20 through 3-23, 3-33, D-4

I

infiltration, 2-2, 2-5, 2-9, 2-10, 2-14, 3-20, 3-36, 4-6, 4-7, 5-2, 5-26, B-4, C-2, C-4
inner security zone, 2-23
insurgency, v, 1-1, 1-2, 1-6, 1-7, 2-1 through 2-8, 2-13, 2-15, 2-16, 2-18 through 2-21, 2-27, 3-2 through 3-4, 3-7, 3-10, 3-12, 3-21, 3-23, 3-38, 4-3, 5-3, 5-26, C-6, D-4, D-5, E-2, E-4 through E-7, E-13, F-7
intelligence cell, 2-16 through 2-18, 3-41
interagency support, 4-1
interdiction, 2-13, 3-16, 3-27 through 3-31, 3-35, 3-36
intergovernmental organizations, 2-10, 2-13

J

joint operation planning process, 5-3
joint special operations area, 1-1, 1-2, 2-8, 2-12, 2-14, 3-19, 3-31, 3-33, 4-3, 4-4, 4-6 through 4-8, 5-5, 5-9, 5-26, A-1 through A-5

L

legal considerations, 4-1, 4-11
linkup, 1-4, 2-9, 2-12, 2-14, 3-22, 3-33 through 3-37, 5-2
logistic support, 4-4, 5-13

M

Marine Corps Forces Special Operations Command, 1-3, A-5
Military Information Support Operations, 1-6, 1-9, 2-9, 2-10, 2-13 through 2-15, 3-8, 3-23, 3-33, 4-1, 4-2, 5-25, 5-26, 5-27, A-4, B-3
mission support site, vi, 3-24, 3-25, 3-27

N

Naval Special Warfare Command, 1-3, A-2, A-3
nonconventional assisted recovery, 3-18
nongovernmental organizations, 2-10, 2-13, 2-27, F-18

O

operations cell, 2-16, 2-17, 4-7
other government agencies, 3-18, 4-2, 5-22
outer security zone, 2-23

P

parallel cells, 2-18
phases of unconventional warfare, 2-8
pilot team, vi, 2-9, 2-10, 4-3
preparation of the environment, 2-8, 3-1, 3-23, 3-39, 4-13, 5-2, 5-6, 5-25, A-3
propaganda, 2-2, 2-4, 2-6, 2-19, 2-24, 2-27, 3-4, 3-12, 3-21, 5-10, C-1, C-6, D-1, D-2, E-3, E-7, E-9
Psychological Operations, 1-9, 3-2, C-8

R

raid, 3-23, through 3-28, 3-31, 3-35, 3-36, D-3, E-8
rear area security, 3-34, 3-37
resupply, 2-12, 4-4 through 4-9, 4-11, A-2 through A-5
riots, 2-21, 3-3, 3-10, 3-20

Index

S

sabotage, v, 2-6, 2-7, 2-12, 2-14, 2-16, 3-8, 3-14 through 3-17, 3-23, 3-28, 3-40, 3-41, C-3, C-4, D-3, D-5, E-9, F-16

shadow government, vi, 2-10, 2-13, 2-21, 4-2, 4-3, 5-27, C-1

subversive activities, 2-6, 2-7

support networks, 2-6, 2-20, 2-24, B-3

T

Title 10, USC, 1-3, 4-9, 4-10

transition, 1-2, 1-3, 2-6, 2-8, 2-14, 2-15, 2-19, 3-35, 3-37, 4-2, 4-3, 5-3, 5-7, 5-8, 5-24, 5-25, 5-30, A-2 through A-4, F-3

U

U.S. Army Special Operations Command, 1-3, 4-10, 5-27, A-1, A-2

U.S. Special Operations Command, v, 1-3, 3-1, 3-19, 4-10, 4-11, 4-13, A-1, A-2, A-4, A-5

unconventional assisted recovery, 3-18, 3-19

underground, 1-1, 1-4, 1-6, 1-7, 2-2, 2-12, 2-16 through 2-21, 2-24, 2-27, 3-2, 3-3, 3-6, 3-8, 3-10, 3-11, 3-13 through 3-17, 3-19, 3-20, 3-23, 3-28, 3-29, 3-31, 3-32, 3-35 through 3-37, 3-39 through 3-42, 4-8, 5-3, A-2 through A-4, C-1, C-5 through C-8, D-1 through D-4, E-4, E-8, F-8

ATP 3-05.1
6 September 2013

By order of the Secretary of the Army:

RAYMOND T. ODIERNO
General, United States Army
Chief of Staff

Official:

GERALD B. O'KEEFE
Administrative Assistant to the
Secretary of the Army
1321201

DISTRIBUTION:
Active Army, Army National Guard, and United States Army Reserve: Not to be distributed; electronic media only.

PIN: 103650-000

www.ingramcontent.com/pod-product-compliance
Lightning Source LLC
Chambersburg PA
CBHW050055230526
45470CB00004B/1546